ROUTLEDGE LIBRARY EDITIONS:
GEOLOGY

T0203807

Volume 17

GEOMORPHOLOGY IN ARID REGIONS

GEOMORPHOLOGY IN ARID REGIONS

Binghamton Geomorphology Symposium 8

Edited by
DONALD O. DOEHRING

Routledge
Taylor & Francis Group

LONDON AND NEW YORK

First published in 1977
First published in 1980 by George Allen & Unwin Ltd

This edition first published in 2020
by Routledge
2 Park Square, Milton Park, Abingdon, Oxon OX14 4RN

and by Routledge
52 Vanderbilt Avenue, New York, NY 10017

Routledge is an imprint of the Taylor & Francis Group, an informa business

© 1977, 1980 Donald O. Doehring

British Library Cataloguing in Publication Data
A catalogue record for this book is available from the British Library

ISBN: 978-0-367-18559-6 (Set)
ISBN: 978-0-429-19681-2 (Set) (ebk)
ISBN: 978-0-367-28020-8 (Volume 17) (hbk)
ISBN: 978-0-367-28022-2 (Volume 17) (pbk)
ISBN: 978-0-429-29923-0 (Volume 17) (ebk)

Publisher's Note
The publisher has gone to great lengths to ensure the quality of this reprint but points out that some imperfections in the original copies may be apparent.

Disclaimer
The publisher has made every effort to trace copyright holders and would welcome correspondence from those they have been unable to trace.

GEOMORPHOLOGY IN ARID REGIONS

DONALD O. DOEHRING,

Editor

A Proceedings Volume of the
Eighth Annual Geomorphology Symposium
held at the State University of New York
at Binghamton, September 23-24,
1977

London
GEORGE ALLEN & UNWIN
Boston   Sydney

Cover photograph of a group of inselbergs near the Piute Mountains, eastern Mojave Desert. Courtesy of William B. Bull, University of Arizona.

First published in 1977
First published by Allen & Unwin in 1980

GEORGE ALLEN & UNWIN LTD
40 Museum Street, London WC1A 1LU

British Library Cataloguing in Publication Data

Geomorphology Symposium, *8th, State University of New York, 1977*
   Geomorphology in arid regions. – (The Binghamton symposia in geomorphology: international series, no. 8).
   1. Geomorphology – Arid regions – Congresses
   I. Title   II. Doehring, Donald O
   III. Series
   551.4   GB611   80-41450

ISBN 0-04-551041-5

Cover photograph of a group of inselbergs near the Piute Mountains, eastern Mojave Desert. Courtesy of William B. Bull, University of Arizona.

Printed in the United States of America

August, 1977

CONTENTS

INTRODUCTION

This monograph is the proceedings volume of the Eighth Annual Geomorphology Symposium, which was held at the Department of Geological Sciences, State University of New York at Binghamton on September 23-24, 1977. The symposium topic, Geomorphology in Arid Regions, uses the term "arid" loosely to include studies from climes which might otherwise be considered semi-arid. This was done to provide a diversity of papers dealing with important problems of interest to geomorphologists today.

The volume begins with "Tex" Reeves' overview and summary of geomorphic studies of intermontane basins. Attention then turns to the piedmont with Steve Wells' report on his alluvial fan research and a contribution on the long standing "pediment problem" by John Moss. Ted Oberlander provides an interesting and intriguing study of slope morphology genesis on the Colorado Plateau. Bill Bull and Leslie McFadden contribute a timely and much needed study of the tectonic geomorphology of a portion of the Mojave Desert.

The next group of papers deals with erosion, sedimentation and predominately fluvial processes. Asher Schick's study of a watershed in the Negev provides an insight into the difficulties and useful techniques of determining sediment budgets in arid regions. Ian Campbell's paper deals with sediment origin as well as transport rates in a semiarid setting. Bill Emmett and Luna Leopold compare empirical measures of sediment discharge with those derived by computations. The geomorphic changes resulting from catastrophic rainfall is the subject of Pete Patton and Vic Baker's report. Bob Curry's paper on the establishment of vigil networks provides important information on field methods relevant to the geomorphic assessment of environmental impact.

The last three papers constitute a change from fluvial

3

topics. Larry Lattman analyzes and reports on the processes and products of caliche weathering in southern Nevada. Jack McCauley, Maurice Grolier and Carol Breed's comprehensive treatment of yardangs is worldwide in scope and includes new field as well as experimental data. Last, but certainly not least, Bill Melhorn and Dennis Trexler supply a paper on their work with eolian processes in the Great Basin. Due to circumstances beyond the control of the authors and editor, three of the above works are represented here by abstracts. Surely the complete papers will be published in prominent places in the near future.

It is apropos, although fortuitous, that this symposium be held in 1977 for it serves to commemorate the hundredth anniversary of G. K. Gilbert's, "Report on the Geology of the Henry Mountains." His report contains, among other things, the first description of a pediment, exposition of the base level concept and his dynamic equilibrium model. In the editor's opinion, this work constitutes the most important single contribution to geomorphology, a view which I am sure is shared by others. Writing in the proceedings volume for the 1975 Binghamton Symposium, Higgins (p. 5) states; "Thus if one were to assign a specific date for the beginning of the modern era in geomorphology, one would most likely choose 1877, the year of publication of T. H. Huxley's "Physiography" and of G. K. Gilbert's chapter on "Land Sculpture" in his report on the geology of the Henry Mountains."

ACKNOWLEDGMENTS

I wish to thank Marie Morisawa and Don Coates for extending the invitation to organize this year's symposium to me. Only after one has done it, can they fully appreciate the difficulty of the task and the magnitude of the service rendered each year by Don or Marie. Especially difficult is the matter of keeping the cost of the proceedings volume down during this time of rising prices and yet having the book available at the time of the meeting. In this regard, my gratitude goes to the authors. For the most part, they adhered to my timetable and provided text requiring relatively little modification.

The entire manuscript for the book was typed (and sometimes edited) by Mrs. Debbie Patterson Tamlin. My wife, Barbara, helped with the proofreading, and printing was done by Citizen Printing Company of Fort Collins, Colorado. My sincere thanks to all who helped.

INTERMONTANE BASINS OF THE ARID WESTERN UNITED STATES

C.C. Reeves, Jr.
Department of Geosciences
Texas Tech University

ABSTRACT

The arid western United States is the type area for
asin and Range topography produced by alternating horst-
graben structure or tilted fault block mountains. Early
investigators such as Gilbert, Powell and Dutton, followed
by Davis and Blackwelder, successfully unraveled the complex
structural and geomorphic history of the Basin and Range
Province by studying the mountains, the pediments and the
bajadas. Little attention was paid to the lowland playa
areas.

Recent studies of intermontane basins, utilizing seismic methods, drill holes and electronic enhancement of sate-
llite imagery, reveals that the basins are unusually complex
tectonic elements, parts of which have subsided (and in many
cases are still subsiding) over considerable periods of geologic time. Multiple, near-parallel faults often border the
grabens, the main parts of which may be displaced by intersecting fault trends. The sedimentary history, particularly
of pre-late Pleistocene sections, is complex and lithologic
distribution in the basins often fails to comply with "established rules".

The common occurrence in intermontane basins of potable
ground water, commercially important evaporite minerals, and
uranium deposits as well as other exotic minerals, insures
an ever-increasing rate of geologic study and evaluation for
the immediate future.

INTRODUCTION

Intermontane basins, or the lowland areas between near-
parallel fault-block mountain ranges, have received, in re-

lation to the vast amount of area comprising such features, relatively little geologic attention. Early workers, such as King (1870), Gilbert (1874, 1875), Powell (1877), Dutton (1880) and Davis (1903, 1905a, 1905b, 1913, 1925) were preoccupied with the origin(s) of the mountain ranges and whether mountain fronts represented fault scarps or fault-line scarps. More recent studies (Eckis, 1928; Blissenbach, 1954; Lustig, 1964; Bull, 1964; Denny, 1965) concentrated on formation of alluvial fans, or pediments (McGee, 1897; King, 1949, 1962; Tator, 1952, 1953; Denny, 1967). During the last 20 years several substantial investigations have dealt with development of Basin-Range structure utilizing modern technological methods (Donath, 1962; Shawe, 1965; Stewart, 1971; Thompson and Burke, 1973; Thompson and others, 1967), thus a great deal of data of unusual importance to mineral explorationists is now available.

Early students of the Basin and Range did not have the finances, instrumentation or drill/coring equipment necessary to study the basins, they thus directed their efforts to the relatively accessible mountain ranges or engaged in dangerous extrapolation and speculation regarding basin origins (Spurr, 1901; Keyes, 1912). Although the financial picture has not necessarily improved over the last 50 years, the use of gravity, magnetics and seismic refraction, and data secured by oil/gas tests, water wells and mineral exploration test holes, now reveal enough detail to allow reasonable paleogeomorphic extrapolations. This study is therefore not a review of the arguments for or against any cause of pedimentation or a discussion of the how, when, and why of mountain front formation in the Basin and Range. Rather, attention is focused on the playa areas of the intermontane basins, to what has geologically happened in the basins, and what probable bedrock configurations can be expected beneath basin fills. That this is needed is indicated by both the general absence of such papers in literature and by the simplified cross-sections often found in textbooks which supposedly illustrate intermontane basin configurations.

The geomorphic development of intermontane basins was outlined by Davis (1904, 1905a) and in retrospect, survived with substantial revisions (Blackwelder, 1928a) into the 1960's (Lobeck, 1939; Thornbury, 1969.). Expectantly, Davis' arid geomorphic cycle, like his humid geomorphic cycle, underwent extensive refinement, simplification, quantification, and criticism. However, Davis (1905a) early mentioned the complexities of Basin and Range structure caused by intermittent faulting, uplift and structural tilting. Thus, any simplifications which later developed were not the result of Davis' arid cycle proposal but resulted from misinterpretations or misjudgements, and thus much of the criticism directed toward Davis' work (Tuan, 1959) was misspent.

The Basin and Range, as classically defined, is an irregularly shaped geographic entity being widest along the Mexican border, narrowing to only a few hundred miles at the latitude of Owens Valley, California, and then spreading across Nevada and into northern Utah (Figure 1). The northern border is uncertain, being masked by the flows of the Columbia Plateau (Figure 1). Although most of the data for this study comes from the basin and Range, many of the conclusions are apropos, in one way or another, to intermontane basins in adjacent physiographic provinces (Figure 1) or to such basins on other continents.

SURFACE GEOLOGY

Intermontane basins appear similar as far as surface expression is concerned. Topographically the intermontane basin or bolson (Tolman, 1909) extends from the divide of one block mountain to the divide of the adjacent mountain mass (Figure 2). In this study the term playa basin refers only to that part of the topographically low area between the outermost, aggradational surfaces of adjacent bajada deposits (Figure 2). The term graben, however, will refer to that part of the basin prescribed by the outermost fringing faults, thus usually describing a feature larger than the playa.

9

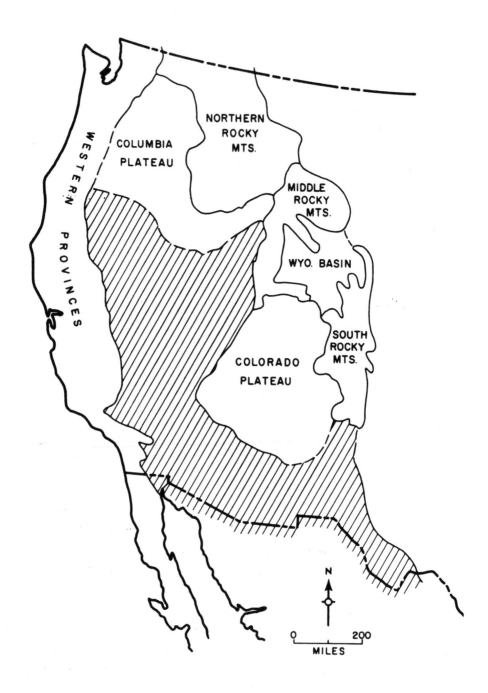

Figure 1. The Basin and Range physiographic province and other arid physiographic provinces in the western United States which contain intermontane basins.

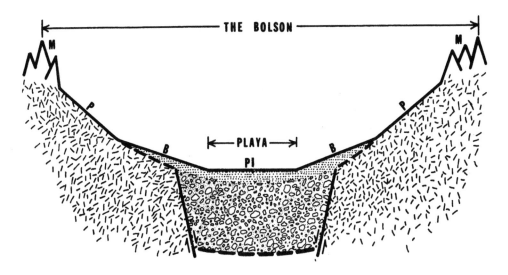

Figure 2. *Generalized cross-section of Basin and Range structure to show limits of the bolson and intermontane basin M = mountains, P = pediment, B = bajada, P1 = playa. Vertical scale greatly exaggerated.*

The outer boundaries of the bajadas are easily determined from topographic maps or field reconnaissance when the pediments have bedrock outcrops sloping steeply (> 5%) into the basin. But in most cases the pediments seldom have over a degree or two of slope, particularly at their outer boundaries, and often imperceptibly grade into the aggregational bajadas. The bajada, or outer area of the playa basin is often surficially similar to the pediment, but cross-sections in drainage channels will show only a veneer of soil and gravel on the outer parts of the pediment. The bajada, however, will consist of wedges of clastic debris and, in places, becomes hummocky. No bedrock is exposed, often being buried under tens of feet of debris.

The central or at least the lowermost area of intermontane basin is occupied by the playa. The term playa, although incorrectly applied to the *barrial* of the intermontane basin, nevertheless is now so entrenched in American geologic literature that it should be accepted. The playa represents the lowest part of the basin and, as such, is frequently flooded by desert downpours or runoff from adja-

11

cent highlands. Sediments consist predominantly of col-
loids, clays and evaporites such as halite, gypsum and sod-
ium sulfate. Surficially the playa exhibits micro-relief
caused by mud cracks, playa grooves, spring mounds, pressure
ridges and sinks (Reeves, 1968), and often parts of the surface will be covered by sand dunes. In recent years numerous playa investigations have been conducted (Neal, 1965;
Reeves, 1968; Motts, 1970) thus playa morphology is well
understood.

In some basins discontinuous troughs develop parallel
to the edge of the basin due to differential movement and
tilting along the basin-fringing faults (Figure 3). From
th air these troughs tend to suggest solution along the
iault planes, however, ground studies always find evidence
of movement which, in many cases, causes a dip reversal of
the lacustrine sediments.

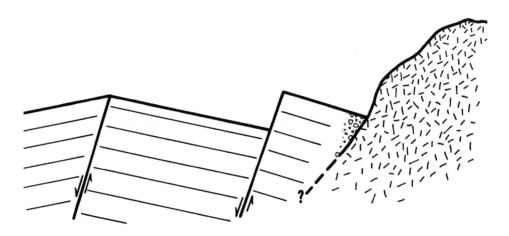

Figure 3. Reverse dip of lacustrine beds along the basin margin produced by differential movements along the fringing faults.

SUBSURFACE GEOLOGY

Structure

The classic, generalized cross-section of an intermontane basin is shown by Figure 2. Intermontane basins are
usually considered by most geologists as simple grabens
which long ago subsided (between single fringing faults)

12

and filled with clastic debris from adjacent highlands.
Many studies (Donath, 1962; Shawe, 1965; Stewart, 1971) show
that this has seldom been the case. Early surface studies
of fringing block mountains revealed Laramide faults, termed
transverse, cross, or tear, which intersected the principal
fault directions (as, for example, in the Muddy Mountains,
Spring Mountains, or Goodsprings, Nevada areas or around
Owens Valley, California) and more recent studies (Donath,
1962; Shawe, 1965; Stewart, 1971) find conjugate fault systems of Tertiary age. It is therefore not surprising that
intermontane basins are often compound grabens with some
areas much deeper than other areas, indicating post-Laramide
movement on at least the Laramide faults in the basin.
Meister (1967) for example, using refraction seismic techniques, clearly identified a "graben in the graben" at Dixie
Valley, Nevada, and Donath (1962) found similar displacements beneath Summer Lake Valley, Oregon (Figure 4). Also,
most intermontane basins are fringed on one or both sides by
multiple faults of Cenozoic age (Figure 5) and contain a
variety of sediments of not only unpredictable lithology and
structure, but of unusual thickness and geographic extent.

Grabens, and intermontane basins in particular, are always considered as narrow and elongate, which happens to be
the case in the Basin and Range of Nevada. However, even in
many parts of Nevada, as well as in the other arid areas of
the world, many grabens or intermontane basins are irregularly shaped due to intersecting structural trends (or high
levels of fill). Thus, some basins contain structural
"sinks" (local low areas within the principal graben caused
by an intersecting graben) structural steps in the floor of
the main structure, or a graben in the graben. This possibility, at least for Deep Spring Valley, California, was suggested by Miller (1925) and the subsurface bedrock floor of
Dixie Valley, Nevada, appears as a "...complex asymmetrical
graben..." (Stewart, 1971). Such structural complexity is
apparently not confined to the western United States as
Tricart (1974) mentions that the Santiago Basin of Chile,

13

which trends north-south, is cut by a southwest-northeast
structural trend.

SW   **NE**

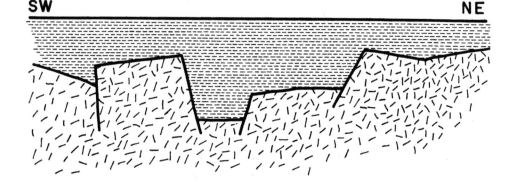

*Figure 4. Seismic-refraction profile from Summer Lake Valley, Oregon
(after Donath, 1962).*

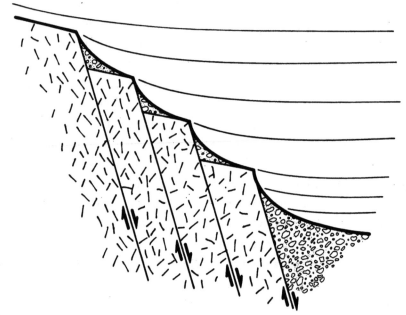

*Figure 5. Schematic illustrating multiple down-to-the-basin faults
typically found bounding grabens in the Basin and Range.*

More often than not intermontane basins are flanked by
more than a single normal fault, although most available
cross-sections illustrate a simplistic structural situation
(Figure 6). The multiple faults are not necessarily paral-

14

lel and often exhibit local and radical changes in strike, but do seem to dip with unusual uniformity. Width of individual fault blocks is also variable, but overall tends to increase in a basinward direction. I suspect that such faulting may commonly extend much farther out into many intermontane basins than is generally expected. This would be particularly true in basins with thick salt deposits which often mask underlying faults, although concealed faults also commonly occur in most large basins regardless of fill lithology.

Many mountain ranges in the western U. S. exhibit rather uninterrupted dip slopes with no fringing fault apparent on the dip slope side. The frontal faults, which are often observable, have produced tilted mountain blocks. Gravity studies in Clark County, Nevada (Kane & Carlson, 1961) indicate several basins represent the in-filled depressed ends of tilted mountain blocks; however, other basins behind tilted mountain blocks are fringed by a typical intermontane basin fault zone. In such cases the fault(s) are usually not observable at the surface due to the rapidity of alluvial sedimentation down the dip slope.

Faulting in the Basin and Range began in Oligocene to Miocene time (Proffett, 1977), however, many fringing faults along intermontane basins continue to be active. Fault scarp displacements along the Sonoma Range, Nevada, occured in 1915, or the scarp in Fairview Valley, Nevada, that formed in 1954, are excellent examples, as are scarps which displace Pleistocene sediments along the East Potrillo Mountains, New Mexico, in Dixie Valley, Nevada (Thompson and Burke, 1973) or in Summer Lake Valley, Oregon (Donath, 1962).

Sediments

Ideally the fringe areas of intermontane basins contain coarser clastics than the more central areas where clays and evaporites are expected. Playas in the central areas of present intermontane basins often exhibit this

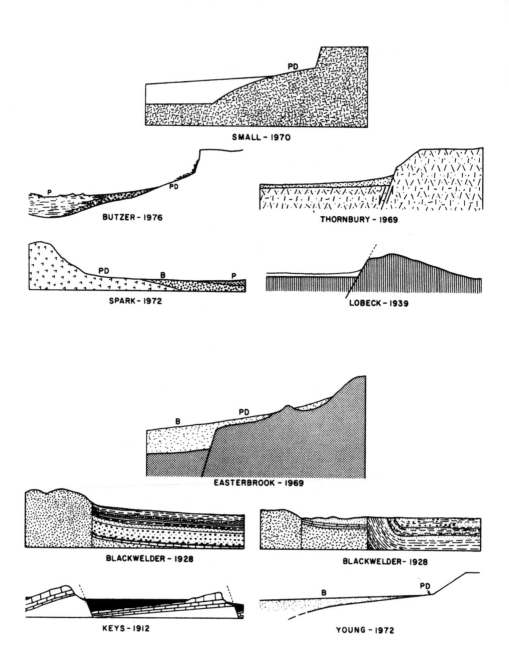

Figure 6. Examples of cross-sections of intermontane basins from present textbooks.

distribution, however, playa positions in most basins have, throughout the geologic history of the basins, often migrated from one to the other side (Figure 7). Thus, in basins which have been occupied with permanent to intermittent lakes, drilling "fences" will often reveal a laterally migrating wedge of lacustrine sediment. Playa shift has been caused by differential movement of the basin and/or the fringing highlands, the playa (or low) moving away from the side experiencing the most rapid uplift at any one time.

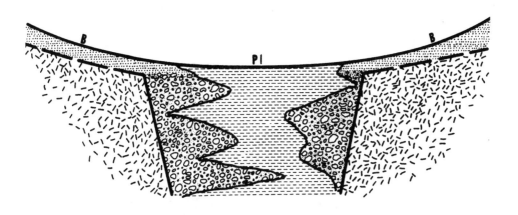

Figure 7. Schematic illustrating the past migrations and widths of the lacustrine wedge in an intermontane basin.

Differential uplift is also illustrated by the migrating wedges of coarse clastic debris. As Tricart (1974) mentions:

> "Very often, the filling up of the basin is therefore asymmetrical, ..."

thus at any one time the alluvial cones extend farther into the basin (and in so doing, push the playa) from the most rapidly rising mountain mass. Interfingering lacustrine carbonates with peripheral gravel beds representative of ancient alluvial fans are well exposed in the Verde Beds of northwestern Arizona (Lehner, 1958).

Coarse clastic deltas often exist where large Tertiary-Quaternary streams entered intermontane basins, however,

17

finer-grained muddy type deltas are also known, indicating streams with low gradients. Fine clastics and even clays frequently exist in near-shore areas, particularly where adjacent mountain masses were either of low relief or composed of hard, non-weathered rock. It is also common in many basins to find both coarse and fine-grained sediments in the near-shore zone. Such basins usually have alluvial fans of early to mid-Cenozoic (?) age buried by early to mid-Cenozoic (?) lacustrine beds located basinward of the principal fringing fault. Later Quaternary lacustrine beds then bury the fault scarp (which often shows recent movement) and overlap the adjacent bajada or pediment (Figure 8).

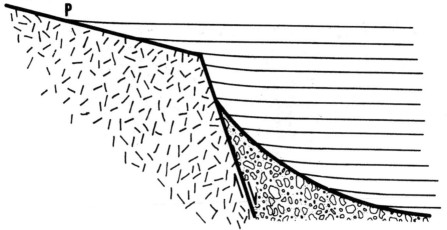

Figure 8. *Schematic relationship of early Cenozoic (?) fan deposits buried by early Cenozoic (?) lacustrine deposits which are then buried by Quaternary lacustrine beds.*

Basin fills, in general, will broadly reflect past climatic conditions as well as surrounding lithology. For example, the deep basin fills in Arizona (Phoenix, Safford, Verde, Haulpai, Detrital Valleys) have thick evaporite sequences (Pierce, 1976; Eaton and others, 1972) compared to deep basin clay fills in northern Nevada or Oregon, whereas shallower fills to the south may consist mainly of sands and gravels compared to clay sections to the north.

It is unusually difficult to correlate older and deeper basin fills. This probably results from different ages for many of the basins as well as being a reflection of local

climate, source, and physiography. Certainly a greater consistency of correlation exists between different basin fills
in the upper few hundred feet of sediment, principally in
very large basins and from the more central parts of the
playas. The uppermost lacustrine sediments are representative of late Pleistocene and Quaternary time and, in many
cases, peripheral carbonates or fossil remains can be dated
by C^{14}. However, once beyond the effective range of C^{14}
there are few reliable stratigraphic markers. In some basin
sediments the base of the Pleistocene section can be recognized, thus the thousands of feet of underlying fill is only
presumed to be of early Cenozoic age.

Problems are also created when basin fills contain massive thicknesses of salts. The salts apparently flow plast-
ically until it becomes difficult to correlate either the
salt horizon or adjacent sediments. Drilling has revealed
displacements of 300 feet in 2 miles.

Depth of fill in intermontane basins can seldom be determined by surface geology. Particularly narrow bolsons may
contain only a veneer or many thousands of feet of fill, and
wide bolsons often have broad areas of shallow cover separated by deep pockets of fill. There also commonly exists a
divergency between the locations of present playas and the
thickest bolson deposits, a situation Mabey (1960) detailed
in the Mojave region.

Depth of fill is best determined by drilling, but few
if any basins have enough drill holes to insure that either
the maximum fill thickness or the configuration of the bedrock bottom is accurately known. In many localities deep
test wells, usually to investigate the possibilities for
development of potable ground water, have yielded substantial information on intermontane basin fill characteristics,
at least to depths of about 1000 feet, but for the most part
dependable control is scarce. Considerable data is available on specific basin fills which have been drilled/cored by
the U.S.G.S. (Smith and Pratt, 1957) and, of course, a
wealth of proprietary data has been collected but is usually

19

not available.

Gravity surveys provide reasonable estimates of fill thickness and basin structure, providing densities used for bedrock, fans and alluvial debris, and playa sediments are "correct" for the basin being studied. An example of a projection based on inaccurate data occurred in the Hueco Bolson near El Paso, Texas, when the U.S.G.S. proposed bedrock at approximately -3400 feet. Total depth on the test well was -4263 feet in lacustrine fill, with a revised bedrock projection of -7000 feet (Cliet, written communication, 1967). Thompson (1959), Kane and Pakiser (1961), Mabey (1960), and Cabaniss (1965) examined the multiple problems associated with determining correct densities. Cabaniss (written communication, 1968)feels a 45 mgal anomaly is usually indicative of about 10,000 feet of fill, thus suggesting a density contrast of about 0.35 gm/cm^3.

Although local deviation exists, the most probable average density contrast between basement rock and basin fills is about 0.5 gm/cm^3. Using gravity, Thompson (1959) delineated 3000 feet of fill in Carson Sink, over 5000 feet in Fairview and Dixie Valleys, 6500 feet near Westgate, and 2000 feet in Edwards Creek Valley, Nevada. Cabaniss (1965) demonstrated up to 12,000 feet of fill in various intermontane basins of California, Kane and Pakiser (1961) found 9000 feet of fill in Owens Valley, California, and Reeves (1969) found up to 13,000 feet of fill in intermontane basins in northwestern Chihuahua, Mexico. Eaton (1939) proposed 16,000 feet of fill in the Ridge Basin, California, composed of Pliocene (13,000) and Pleistocene (3000) lacustrine strata.

Paleogeomorphology

Intermontane basins throughout the western United States, and in most of the world's block faulted regions, exhibit various stages of development, as measured by elevation of adjacent mountain ranges, extent of pediments, depth of fill and multiplicity of faulting. Certain basins,

such as the Shirley Basin at the southern end of the Powder
River Basin, Wyoming, have been intensively studied due to
the occurrence of commercial uranium deposits, but for all
practical purposes, most intermontane basins are geologically unknown.

Many of the range front Cenozoic faults and fault zones
in the Basin and Range now exhibit thousands of feet of
throw, thus the question arises as to probable morphology
during late Tertiary-early Quaternary time. Pre-Cenozoic
morphology is particularly conjectural in local areas although the regional pattern can be surmised from sporadic
outcrops. Throughout Nevada outcrops of Paleozoic-Mesozoic
rocks exhibit north-south fold axes (Shawe, 1965), while in
the far western part of the Basin and Range fault blocks
trend northeast to northwest (Gilbert and Reynolds, 1973).
However, regardless of trend of the positive features early
Cenozoic lacustrine deposits must have accumulated in associated Cenozoic intermontane basins. These are the deposits
which have been drilled and indicated by gravity beneath recognized Plio-Pleistocene fills and which occur beneath present mountain masses as at the Clan Alpine Range, Nevada
(Meister, 1967). Reversals of topography have also occurred
in some areas (Gilbert and Reynolds, 1973) in that early
Cenozoic high areas later (Miocene-Pliocene) received sedimentary fills and early Cenozoic lows now exist beneath
present mountain ranges.

The principal question concerning paleogeomorphology of
the Basin and Range regards the relative rates of uplift,
tilting, and subsidence. Lithofacies studies prepared for
paleoenvironmental investigations of several basins indicates that uplift and subsidence have been continually and
rather uniformly active throughout the sedimentary history
of the basins. However, adjacent uplifts have often pro-
ceeded at varying rates. Exaggerated relief or relief
greater than that presently found is not indicated in the
sedimentary sections, thus I suspect that the Basin and
Range today exhibits what might be termed "maximum geomorphic development".

REFERENCES CITED

Blackwelder, E., 1928a, The recognition of fault scarps:
Jour. Geol., v. 36, p. 289-311.

_____ , 1928b, Origin of the desert basins of southwest United States: Geol. Soc. America, Bull., v. 39,
p. 262-263.

Blissenback, E., 1954, Geology of alluvial fans in semi-arid
regions: Geol. Soc. America, Bull., v. 65, p. 175-190.

Bull, W. B., 1964, Geomorphology of segmented alluvial fans
in western Fresno County, Calif.:   U. S. Geol. Survey
Prof. Paper 352-E, 128 p.

Butzer, K. W., 1976, Geomorphology from the Earth: Harper
& Row, New York, 463 p.

Cabaniss, G. H., 1965, Geophysical studies of playa basins;
in, Geology, mineralogy and hydrology of U. S. playas:
Envir. Res. Paper 96, U.S. Air Force Cambridge Res.
Labs, p. 123-147.

Davis, W. M., 1903, The mountain ranges of the Great Basin:
Harvard Mus. Comp. Zoology Bull., v. 42, p. 129-177.

_____, 1905a, The geographical cycle in an arid cli-
mate: Jour. Geol., v. 13, p. 381-407.

_____, 1905b, The Wasatch, Canyon and House ranges,
Utah: Harvard Mus. Comp. Zoology Bull., v. 49, p. 13-
56.

_____, 1913, Nomenclature of surface forms on faulted
structures: Geol. Soc. America Bull., v. 24, p. 163-
216.

_____, 1925, The Basin Range problem: Proc. Utah
Academy of Sciences, v. 11, p. 387-392:

Denny, C. S., 1965, Alluvial fans in the Death Valley region
Calif. and Nevada: U. S. Geol. Survey, Prof. Paper
466, 62 p.

_____, 1967, Fans and pediments: Am. Jour. Science,
v. 256, p. 81-105.

Donath, F. A., 1962, Analysis of Basin-Range structure,
South-central Oregon: Geol. Soc. America Bull.,
v. 73, p. 1-16.

Dutton, C. E., 1880, Geology of the high plateaus of Utah: U. S. Geog. and Geol. Survey Rocky Mt. Region (Powell), G.P.O., 307 p.

Easterbrook, D. J., 1969, Principles of Geomorphology: McGraw-Hill, New York, 462 p.

Eaton, G. P., Peterson D. L. and Schumann H. H., 1972, Geophysical geohydrological, and geochemical reconnaissance of the Luke Salt Body, central Arizona: U. S. Geol. Survey, Prof. Paper 753, 28 p.

Eaton, J. E., 1939, Ridge Basin California: Am. Assoc. Petrol. Geol. Bull., v. 23, p. 517-558.

Eckis, R., 1928, Alluvial fans of the Cucamonga district, southern Calif.: Jour. Geol., v. 36, p. 224-247.

Gilbert, C. M. and M. W. Reynolds, 1973, Character and chronology of basin development, western margin of the Basin and Range Province: Geol. Soc. America Bull., v. 84, p. 2489-2510.

Gilbert, G. K., 1874, Preliminary geologic report, expedition of 1872: U. S. Geog. and Geol. Survey west of the One Hundredth Meridian (Wheeler), Prog. Report, G.P.O., p. 48-52.

_____, 1875, Report on the geology of portions of Nevada, Utah, California, and Arizona examined in the years 1871 and 1872: Rpt., U. S. Geog. and Geol. Surveys west of the One Hundredth Meridian (Wheeler), G.P.O., No. 3, p. 17-187.

Kane, M. F., and Pakiser L. D., 1961, Geophysical study of subsurface structure in southern Owens Valley, California: Geophysics, v. 26, p. 12-26.

_____, and Carlson J. E., 1961, Gravity anomalies isostasy and geologic structure in Clark County, Nevada: U. S. Geol. Survey, Prof. Paper 424-D.

Keyes, C. R., 1912, Deflative scheme of the geographic cycle in an arid climate: Geol. Soc. America Bull., v. 23, p. 537-562.

King, C. E., 1870, Systematic geology: Rpt. U. S. Geol. Explor. 40th parallel: G.P.O., 803 p.

King, L. C., 1949, The pediment landform: some current problems: Geol. Mag., v. 86, p. 245-250.

_____, 1962, Morphology of the Earth: Oliver and Boyd, Edinburgh, 699 p.

Lehner, R. E., 1958, Geology of the Clarkdale Quadrangle, Arizona: U. S. Geol. Survey, Bull. 1021-N, p. 511-592.

Lobeck, A. K., 1939, Geomorphology - an introduction to the study of landforms: McGraw-Hill, New York, 731 p.

Lustig, L. K., 1964, Clastic sedimentation in Deep Springs Valley, Calif.: U. S. Geol. Survey, Prof. Paper 352F, 192 p.

McGee, W. J., 1897, Sheetflood erosion: Geol. Soc. America Bull., v. 8, p. 87-112.

Mabey, D. R., 1960, Gravity survey of the western Mojave Desert, Calif.: U. S. Geol. Survey, Prof. Paper 316D.

Meister, L. J., 1967, Seismic refraction study of Dixie Valley, Nevada: in, Geophysical Study of Basin-Range structure, Dixie Valley Region, Nevada: U.S. Air Force Cambridge Res. Labs, Report 66-848, Pt. 1, 72 p.

Miller, W. J., 1925, Geology of Deep Spring Valley, California: Geol. Soc. America Bull., v. 36, p. 510-525.

Motts, W. S., (ed), 1970, Geology and hydrology of selected playas in western United States: U.S. Air Force Cambridge Res. Labs, 286 p.

Neal, J. T., (ed), 1965, Geology, mineralogy and hydrology of U. S. playas: U.S. Air Force Cambridge Res. Labs, 176 p.

Pierce, H. W., 1976, Tectonic significance of Basin and Range thick evaporite deposits: Ariz. Geol. Soc. Digest, v. 10, p. 325-339.

Powell, J. W., 1877, Rpt. Geog. and Geol. Survey Rocky Mt. Region: G.P.O., 19 p.

Proffett, J. M., Jr., 1977, Cenozoic Geology of the Yerington district, Nevada, and implications for the nature and origin of Basin and Range faulting: Geol. Soc. America Bull., v. 88, p. 247-266.

Reeves, C. C., Jr., 1968, Introduction to Paleolimnology: Elsevier Pub. Co., Amsterdam, 228 p.

_____, 1969, Pluvial Lake Palomas, Northwestern Chihuahua, Mexico: in, 20th ann. Field Conference Guidebook of the Border Region, New Mexico Geol. Soc., p. 143-154.

Shawe, D. R., 1965, Strike-slip control of Basin-Range structure indicated by historical faults in western Nevada: Geol. Soc. America Bull., v. 76, p. 1361-1378.

Small, R. J., 1970, The Study of Landforms: Cambridge Univ. Press, 486 p.

Smith, G. I. and Pratt, W. P., 1957, Core logs from Owens, China, Searles, and Panamint Basins, Calif.: U. S. Geol. Survey, Bull. 1045-A, 62 p.

Sparks, B. W., 1972, Geomorphology: Longman Group, London, 530 p.

Spurr, J. E., 1901, Origin and Structure of the Basin Ranges: Geol. Soc. America Bull., v. 12, p. 217-270.

Stewart, J. H., 1971, Basin and Range structure: a system of horsts and grabens produced by deep-seated extension: Geol. Soc. America Bull., v. 82, p. 1019-1044.

Tator, B. A., 1952, Pediment interstream surfaces of the Colorado Springs Region: Geol. Soc. America Bull., v. 63, p. 355-374.

_____, 1953, Pediment characteristic and terminology: Ann. Assoc. Am. Geog., v. 43, p. 47-53.

Thompson, G. A., 1959, Gravity measurements between Hazen and Austin, Nevada - a study of Basin and Range structure: Jour. Geophys. Res., v. 64, p. 217-229.

_____, L. J. Meister, A. T. Herring, T. E. Smith, D. B. Burke, R. L. Kovach, R. O. Burford, I. A. Salehi and M. D. Wood, 1967, Geophysical Study of Basin-Range Structure, Dixie Valley, Region, Nevada: U.S. Air Force Cambridge Res. Labs, Report 66-848, 286 p.

_____, and D. B. Burke, 1973, Rate and direction of spreading in Dixie Valley, Basin and Range Province, Nevada: Geol. Soc. America Bull., v. 84, p. 627-632.

Thornbury, W. D., 1969, Principles of Geomorphology: John Wiley & Sons, New York, 618 p.

Tolman, C. R., 1909, Erosion and Depositions in southern Arizona bolson region: Jour. Geol., v. 17, p. 136-163.

Tricart, J., 1974, Structural Geomorphology: translated by S. H. Beaver and E. Derbyshire: Longman Group, Ltd., London, 305 p.

Tuan, Y. F., 1959, Pediments in southeastern Arizona: Univ. Calif. Pub. in Geography, v. 13, 162 p.

Young, A., 1972, Slopes: Oliver and Boyd, Edenburgh, 288 p.

GEOMORPHIC CONTROLS OF ALLUVIAL FAN DEPOSITION IN THE SONORAN DESERT, SOUTHWESTERN ARIZONA

Steve G. Wells

Department of Geology
University of New Mexico

ABSTRACT

Processes of alluvial fan aggradation in southwestern Arizona were analyzed for 12 drainage basins (1-300 km^2) and their associated ephemeral washes. Areas of net deposition of sediments along ephemeral washes may occur locally in incised reaches and extensively in reaches that are not incised below the fan surface. Incised reaches of washes contain topographically low channels floored with coarse grained sediments. These low channels are flanked by berms of fine grained sediment. Deposition in incised reaches occurs by lateral extension of coarse channel sediments over berm sediments producing a local maximum channel width. The width of the channel where this deposition occurs is proportional to the distance from the headwaters divide of the wash (L_c) and the drainage area (A_d) above the maximum channel width. The distance from the head of the wash and drainage area above the maximum channel width are related by the expression, $A_d = 2.5 \, L_c^{1.1}$.

Aggradation outside the low channels occurs as deposition of fine grained berm sediments. These sediments are deposited as water is lost by infiltration into coarse channel sediments. Berm sediments extend laterally and coalesce over low interfluves in distal regions of the fans. The drainage area (A_d) and wash length from the drainage divide (L_b) above this berm aggradation are related by the power function, $A_d = 0.138 \, L_b^2$. The source area of sediment provided for berm aggradation is the interfluves in the upper drainage basin. Berm aggradation is directly related to higher values of mean interfluve width of a basin, or larger source areas.

All ephemeral washes and drainage basins in southwestern Arizona which display channel deposition and berm aggradation obey the two mathematical expressions given. Thus, these two power functions may be used for predicting areas of net deposition and associated sediment types on alluvial fans, by relating physiographic characteristics of drainage basins to clastic deposition in southwestern Arizona.

INTRODUCTION

Periods of alluvial fan building in desert regions of southwestern United States have been attributed to changes in the tectonic setting (Bull, 1964) or climatic setting (Lustig, 1965; Melton, 1965). Changes in these factors may induce increased sediment yield and long periods of sedimentation along desert washes (Denny, 1965). Climate and tectonism may be considered as independent variables which may initiate fan building episodes if the alluvial fan system is evaluated as a whole and is studied over its entire geologic history. However, in searching for those specific variables which cause aggradation along washes, and subsequent fan building, shorter time spans and components of the alluvial fan system must be considered. Variables such as hydrology and drainage network morphology of alluvial fans become significant when considering recent fan building processes. Study of the relative importance of time, space, and causality has been emphasized for erosional landscapes and processes (Schumm and Lichty, 1965), but a similar approach for depositional landscapes has received little or no attention.

It is the primary purpose of the paper to determine those geomorphic variables which influence Holocene alluvial fan aggradation in the Sonoran Desert of Arizona (Figure 1). A secondary purpose is to demonstrate that these variables operate in an orderly fashion across alluvial fans so that areas of net aggradation may be quantitatively determined.

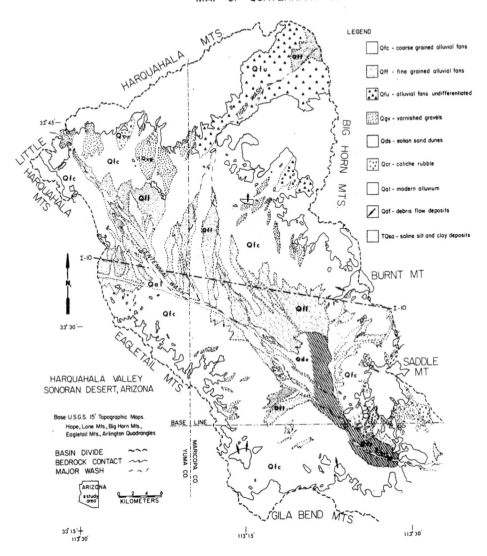

MAP OF QUATERNARY SURFICIAL DEPOSITS

LEGEND

Qfc - coarse grained alluvial fans

Qff - fine grained alluvial fans

Qfu - alluvial fans undifferentiated

Qgv - varnished gravels

Qds - eolian sand dunes

Qcr - caliche rubble

Qal - modern alluvium

Qdf - debris flow deposits

TQsa - saline silt and clay deposits

HARQUAHALA VALLEY
SONORAN DESERT, ARIZONA

Base U.S.G.S. 15' Topographic Maps.
Hope, Lone Mts., Big Horn Mts.,
Eagletail Mts., Arlington Quadrangles

BASIN DIVIDE
BEDROCK CONTACT
MAJOR WASH

ARIZONA
study area

0 2 4 6
KILOMETERS

Figure 1. Geomorphic map of the Harquahala Valley, southwestern Arizona.

To achieve these goals, sedimentologic variables (types and distribution of wash sediments), hydrologic variables (runoff and infiltration) and drainage network variables (drainage density) are compared for 12 small watersheds in southwestern Arizona (Figure 2). These subwatersheds are selected from a single larger basin, the Harquahala Valley, which is bounded by six isolated, fault block mountains (Figure 2). These minor watersheds drain the interior of these mountain ranges and the piedmonts. They vary in area from 1 to 285 km^2. Major characteristics of these subwatersheds are summarized in Table 1.

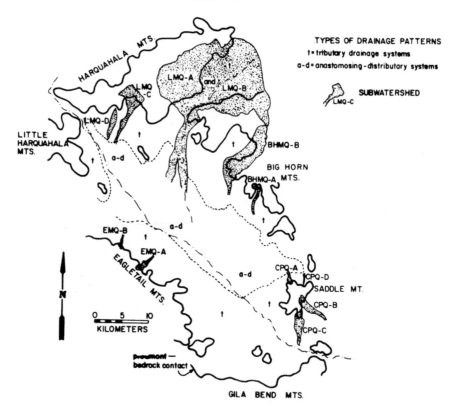

Figure 2. *Location map for subwatersheds and areal extent of drainage patterns in Harquahala Valley. Dashed line represents contact between 2 types of drainage patterns.*

SUB-WATERSHED	TOTAL DRAINAGE AREA (km²)	TOTAL WASH LENGTH* (km)	RATIO OF PIEDMONT TO BEDROCK	MEAN INTERFLUVE WIDTH (W_I) (km)	DOMINANT LITHOLOGIES OF SOURCE AREA	PRESENCE OF BERM AGGRADATION
LMQ-A	290.0	35.5	0.83	(not analyzed due to scale)	Igneous, Metamorphic Sedimentary	Yes
LMQ-B	285.0	33.3	0.79	(not analyzed due to scale)	Igneous, Metamorphic Sedimentary	Yes
LMQ-C	20.4	14.1	1.05	0.185	Metamorphic, Sedimentary	Yes
LMQ-D	3.9	4.9	all piedmont	0.125	Sedimentary, Metamorphic	No
BHMQ-A	4.9	7.0	1.95	0.159	Volcanic	Yes
BHMQ-B	44.5	18.6	0.46	0.313	Volcanic, Igneous	Yes
CPQ-A	2.2	2.8	3.64	0.217	Volcanic	Yes
CPQ-B	11.4	11.4	0.65	0.294	Volcanic	Yes
CPQ-C	4.3	9.0	0.91	0.111	Volcanic	No
CPQ-D	1.0	3.0	all piedmont	0.159	Volcanic	Yes
EMQ-A	3.5	5.8	1.30	0.086	Igneous, Volcanic	No
EMQ-B	1.3	3.0	0.55	0.052	Volcanic	No

* Wash length is measured from the drainage divide unless otherwise stated.

Table 1. Geomorphic characteristics of subwatersheds and their associated principal wash in the Harquahala Valley.

DRAINAGE BASIN CHARACTERISTICS OF THE HARQUAHALA VALLEY

Topography

The Harquahala Valley is an elongate basin which trends N55°W and covers approximately 2000 km^2. The total basin length is 75 km and the average basin width is 20 km. Two thirds of the basin's area is comprised of coalescing piedmont slopes which form the bolson plain. Elevation of the bolson plain ranges from over 600 m to slightly less than 300 m above mean sea level. The remaining one third of the basin's area is developed on bedrock outcrops of mountains and inselbergs. The bolson plain of the Harquahala Valley is flanked by six major mountain ranges which, in order of decreasing maximum relief, are the Harquahala, Big Horn, Eagletail, Little Harquahala, Saddle, and Gila Bend mountains (Figure 1). The two largest ranges, the Harquahala and Big Horn mountains, contribute 60 percent of the total bedrock area.

General Drainage Features

The axial drainage of the Harquahala Valley is the southeasterly flowing Centennial Wash (Figure 1) which is tributary to the Gila River. Centennial Wash originates outside the study area and flows southward through a gap between the Little Harquahala and Harquahala mountains. It leaves the study area through a gap between the Gila Bend and Saddle mountains and joins Gila River. Centennial Wash is deeply incised and contains coarse alluvium where it passes through these gaps. These two gaps are the only places that the axial drainage is a distinct channel (Ross, 1923). In all other places, Centennial Wash is a very broad and shallow feature containing fine alluvium. It is ephemeral and very rarely carries discharge through the entire length of the wash.

The majority of smaller washes originate on the bolson plain, and only a few washes drain from the mountain interiors. Not all the flanking ephemeral washes reach Centennial Wash, but the majority are oriented orthogonal to this

32

axial drainage. Two types of drainage patterns are discernible from color Skylab photography (S190B) at a scale of 1:125,000. Figure 2 is a map showing the areas of these drainage patterns in the Harquahala Valley as mapped from the Skylab photography. These patterns are; 1) deeply incised, tributary drainage systems developed on the upper piedmont slope and; 2) broad, shallow anastomosing drainage systems developed on the lower piedmont. The tributary drainage lines are incised from 1 to 25 m below the level of the bolson plain. The anastomosing distributary drainage lines are incised to a maximum of 0.5 m below the bolson plain. The boundary between these two drainage types is not sharp but is transitional and indicates the place where one drainage pattern type dominates over the other type (Figure 2).

Precipitation and Runoff Relationships

Two major factors affect surface runoff, climate and physiography (Chow, 1964). For an arid region such as Harquahala Valley, precipitation and evaporation are the most important climatic variables, and basin and channel characteristics are important physiographic variables.

Rainfall in the Sonoran Desert can be divided into two types (Dunbier, 1968). Precipitation during the winter months of November through April covers larger areas, is longer in duration per storm and is not as intense as summer precipitation. In the summer months of May through October, the precipitation is more localized, and is of short duration and high intensity. Observations during winter and summer field work indicate that little runoff is produced by winter rainfall; whereas, high surface runoff in the form of flash floods is produced by summer storms. Runoff records at the gauging station on Centennial Wash and precipitation data from the study area are used to determine seasonal variations between precipitation and runoff. Although various factors reduce the correlation of surface runoff and precipitation, the Spearman Rank Correlation Coefficient (r_s) (Siegel, 1956) for summer runoff and precipitation is 0.81 and is significant at $\alpha = 0.05$. The r_s for winter precipitation is 0.50 and is not significant at $\alpha = 0.05$. Therefore, during summer months there is a positive correlation between rainfall and surface runoff in Centennial Wash.

Thus, field observations and statistical analysis indicate seasonal differences in runoff and precipitation relationships. In order to determine the magnitude and frequency of these differences, a comparison is made between the percentage of monthly rainfall and runoff averaged over a ten year period. Fifty-four percent of the precipitation occurs during the winter months and produces 14 percent of the yearly runoff. In the summer months, 46 percent of the precipitation results in 86 percent of the surface runoff. It is concluded that even though rainfall is distributed evenly between winter and summer seasons, over 80 percent of the runoff is associated with intense summer storms of short duration. Dubief (1953) noted similar relationships in the Saharan Desert, where low rainfall with high intensity in the summer produces more runoff than a larger amount of less intensity rainfall in the winter.

STREAMFLOW CHARACTERISTICS OF EPHEMERAL WASHES

In the Harquahala Valley, the only source of runoff is derived from precipitation. All washes are dry except during periods of rainfall, when intense flooding may occur. The highest discharge recorded on Centennial Wash for a ten-year period is 410.5 m^3/sec. and the average discharge is 0.09 m^3/sec. (U.S. Geological Survey, 1970, 1975). Flow in Centennial Wash occurs approximately 5 percent of the time, and the average discharge of 0.09 m^3/sec. is equalled or exceeded only 2.9 percent of the time (Wells, 1976). Only one tributary to Centennial Wash in the Harquahala Valley has been successfully gauged for a period of three years (Wehro, 1967). It is in the northern portion of the basin and has a drainage area of 7.2 km^2.

Flow in this tributary occurs less than three percent of the time and its maximum discharge for a three year period was 20 m^3/sec.

The hydrographs of summer flood events on ephemeral washes have steeply rising limbs, very peaked curves, and less steep, falling limbs. These hydrographs are interpreted as having very short durations of peak flow, which follows very closely behind the front of the flood pulse. The steep, rising limbs on the hydrographs indicate a short rise time. Renard and Keppel (1966) have analyzed the hydrographs of ephemeral streams in southeastern Arizona and central New Mexico. They found that the short, intense rainfall of summer storms causes the largest amount of runoff, and that the processes by which this runoff is transmitted from portions of the basins to a given channel cross-section are reflected in the hydrographs. Hydrographs from runoff in the southern desert of Israel (Schick, 1970) show steep, rising limbs similar to those found in this study and by Renard and Keppel. These investigators have related the short rise time on these hydrographs to transmission losses into the beds of washes.

Measurement of transmission losses in the eastern Sonoran Desert range from 23 to 80 percent (Babcock and Cushing, 1942; Burkham, 1970). Transmission losses cannot be evaluated quantitatively on Centennial Wash since there is only one gauging station. However, transmission losses in the Harquahala Valley must be as high as those described by Babcock and Cushing (1942) and Burkham (1970). Field observations indicate that a summer storm in 1974, which produced approximately 1 cm of rainfall in a two hour period, did not yield enough runoff to reach the axial drainage and exit the Harquahala Valley. Thus, transmission losses, combined with evaporation, affect 100 percent reduction in runoff on the bolson plain.

SEDIMENTOLOGY OF EPHEMERAL WASHES

The principal washes of the subwatersheds can be divided into two reaches based on differences in sedimentology
and topography (Figure 3). The first portion, or channel,
is characterized by coarse sediments, and the topographically higher second portion, or berm is characterized by fine
sediments (Figure 3). Differences between channel and berm
sediments are shown by sieve analysis (Figure 4). The berm
sediments typically have grains whose diameters are less
than 0.125 mm. Channel sediments have grain sizes which are
generally larger than 2 mm in diameter. The topographic
relationship between these two types of reaches is shown in
Figure 3 and is common to most ephemeral washes in the
Harquahala Valley.

W_m = maximum width of flow

W_c = maximum width of channel

W_b = maximum width of berm

D_m = maximum depth of flow

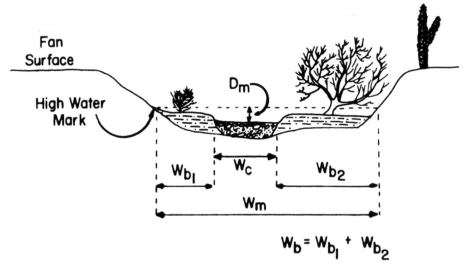

$$W_b = W_{b_1} + W_{b_2}$$

Figure 3.   Idealized cross-section of ephemeral wash showing berm and
channel portions in addition to the hydraulic geometry based
on high water marks from a previous flood. (Note: scour
not indicated.)

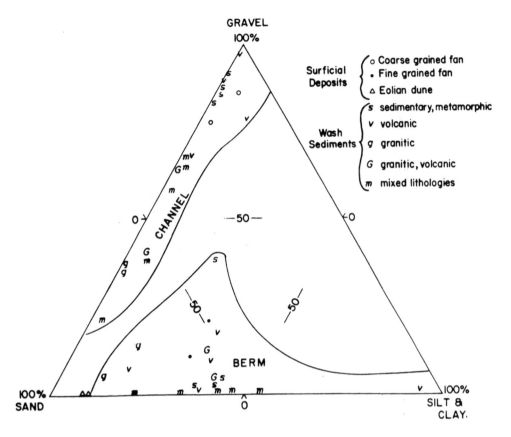

GRAVEL
100%

Surficial Deposits
- o Coarse grained fan
- • Fine grained fan
- △ Eolian dune

Wash Sediments
- s sedimentary, metamorphic
- v volcanic
- g granitic
- G granitic, volcanic
- m mixed lithologies

CHANNEL

BERM

100% SAND

SILT & CLAY.

Figure 4.   Relationship between particle size and lithology of clasts for channel and berm sediments and Quaternary surficial deposits.

The relationship between the lithology and size of the sediment in the berm and channel of ephemeral washes is illustrated in Figure 4.   The difference between the two sediment size populations is obvious except for granitic sediments.   Most berm deposits, composed of detritus derived from sedimentary, metamorphic, volcanic rocks, and combinations of these lithologies, contain approximately 60 percent fine sand and silt, but berms composed of granitic detritus have higher percentages of sand.   Granite, which commonly weathers to a uniform sand size (grus), obscures the difference in particle size typical of berm and channel clasts. Channel deposits usually have a mixture of less than 30 percent sand and more than 70 percent gravel.

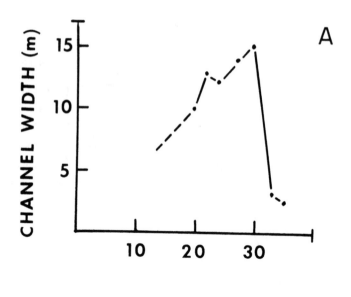

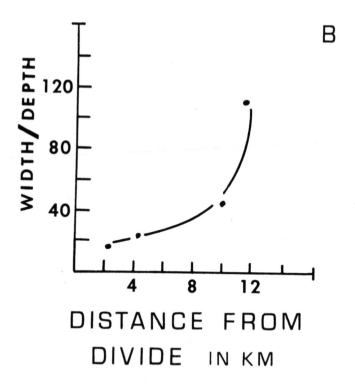

Figure 5. A. *Variation in width of channel downwash from the drainage divide. Note maximum channel width.*

B. *Realtionship between width-depth ratio (based on high water marks) and distance downwash for those minor watersheds which display berm aggradation.*

Downwash Variations in Channel and Berm Sediments

The weight percent of fine grained berm sediments increases downwash from the drainage divide. The maximum width of berms increases with distance downwash; whereas, the width of the channels increases with distance from the drainage divide, passes through a maximum, and then decreases (Figure 5). That is, a maximum width of channel commonly occurs at some point along the course of the wash (Figure 5). These observations suggest a change in competency of runoff downwash. Maximum particle sizes (measured as maximum diameter) decrease with distance downwash in 90 percent of the principal washes. This decrease in maximum particle size supports the decrease in tractive force with distance from the headwaters.

Berm and Channel Deposition

The decrease in tractive force results from the depletion of discharge by transmission losses in the channel sediments. A reduction in discharge results in the overloading of the floodwaters with fine berm sediments as the coarse sediments are deposited. Leopold and Miller (1956), Chow (1964), Bull (1968) and Schick (1970) conclude that suspended load increases faster than the discharge downwash as transmission losses reduce the volume of water. Additional observations reveal that sheetfloods in deserts commonly have high concentrations of fine suspended sediment (McGee, 1897; Joly, 1952; Blissenbach, 1954).

The increase in the width of berms and increase in percent by weight of fines in berm sediments occurs through this overloading process as runoff is depleted by infiltration. Aggradation of berm sediments results during sheetflow and the lateral extension of fine sediments over shallow interfluves in the distal reaches of the piedmont slopes. Thus, the width-depth ratio increases rapidly where berm aggradation occurs and can be seen as a sudden increase in the slope of plots of width-depth ratio versus distance downwash (Figure 5). The coalescing of berm sediments results in local fine grained alluvial fans whose apices occur in the reach of increased width-depth ratios. These fine grained fans represent zones of net aggradation in the distal portions of the subwatersheds. The majority of these small watersheds display berm aggradation (to be discussed in detail later). Aggradation by coalescing berm sediments causes buried vegetation, buried desert pavement surfaces, and buried soil horizons. Net accumulation of channel sediments occurs at the point of local maximum channel width. Inspection of this point on the surface indicates that it is composed of coarser sediment which has been deposited over the finer berm materials. The point of local maximum channel width is not considered aggradational as these sediments do not accumulate for long periods of time due to scouring by floodwaters.

The point of net accumulation of sediments at the maximum channel width usually is upstream of the point at which tributary drainage changes to anastomosing distributary drainage (Figure 2). It is concluded that channel deposition occurs by streamflood (channelized flow) and not sheetflood in these watersheds. During flood events in ephemeral washes in the Sonoran Desert, a maximum discharge occurs at some point along the reach of the wash. The depth of flow increases prior to this maximum discharge, and with increasing depth the shear stresses, or tractive forces, increase. It follows that deposition of the channel sediments and development of the local maximum channel width must occur after the maximum discharge of an ephemeral wash as the depth of flow and tractive forces decrease. Thus, the sediments are unable to be moved by floodwaters and deposition occurs. Berm aggradation and channel aggradation occur during the late stage of a flood event, and both berm aggradation and channel deposition are related to transmission losses. Berm aggradation results from increasing suspended load-discharge ratios, and channel deposition occurs when critical tractive forces are no longer able to move coarse channel sediments.

RELATIONSHIPS BETWEEN WASH PROCESSES AND
GEOMORPHIC PARAMETERS

The hydrologic behavior of the ephemeral washes in
these watersheds is influenced by the wash discharge and
transmission losses. Those geomorphic parameters which are
related to the amount of discharge from a basin and its subsequent depletion are drainage area and wash length. For
both humid and semi-arid regions, it has been shown that
discharge is directly proportional to the drainage area
(Hack, 1957; Leopold, Miller, and Wolman, 1964). Thus, runoff and sediment transport are a function of the size of the
drainage basin. The loss of discharge is proportional to
the length and width of the wash over which infiltration can
occur.

Channel Deposition

Spearman Rank Correlation Coefficients, r_S, are computed for the local maximum channel width and its distance
from the drainage divide for those subwatersheds showing
this type of channel deposition. The value of r_S is 0.38
and is significant at $\alpha = 0.05$ (all correlation coefficients
cited are significant at 0.05 level unless otherwise stated).
This indicates that the length of the channel above the maximum channel width is related to the local maximum width of
the channel. That is, with longer washes, the channel will
be wider at the local maximum. A logarithmic plot of local
maximum channel width and wash length to that point is
given in Figure 6, which supports a positive relationship
between these two variables. The relationship between
local maximum channel width and the drainage area above that
maximum is also tested by the Spearman Rank Correlation
Coefficient. The correlation coefficient is 0.82. A logarithmic plot of the local maximum width of the channel
and the drainage area above that point shows a positive relationship between drainage area and maximum width of the
channel (Figure 6).

41

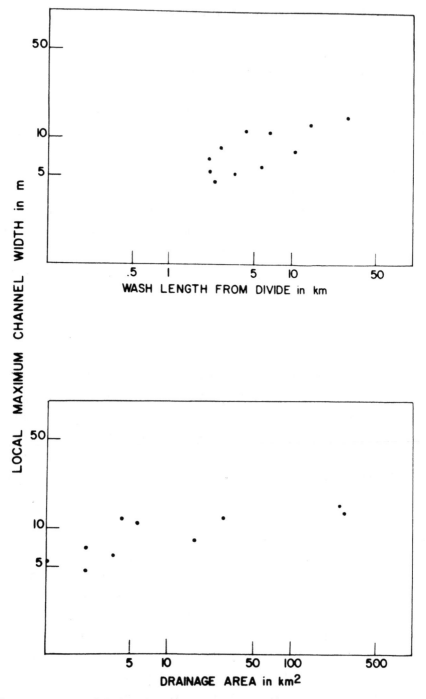

Figure 6. Logarithmic plots of drainage area and wash length above the local maximum channel width or point of channel aggradation.

The local maximum width of the channel in 80 percent of
these basins is proportional to wash length and drainage
area above the local maximum. Specifically, the relationship between the length of the channel and drainage area
of these low order basins at the local maximum channel
width is shown by r_S which is equal to 0.988 and is significant.

A power function regression equation is used to anal-
yze the relationship between the drainage area (A_d) and
wash length (L_c) upwash from the maximum channel width
(Figure 7). The regression equation is, $A_d = 2.5 \ L_c^{1.1}$.

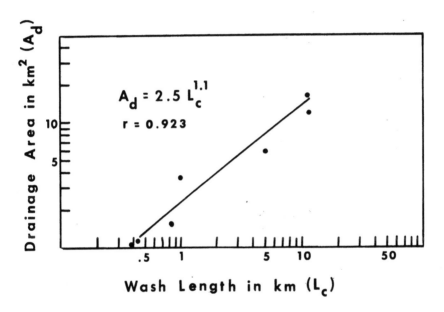

Figure 7. *Relationship between wash length and drainage area above
the local maximum channel width and regression analysis
results.*

Berm Aggradation

A specific drainage area and wash length are related
to berm aggradation. The drainage area and wash length upwash from the apices of the fine grained fans are plotted
on a logarithmic scale in Figure 8. A power function equation relating drainage area (A_d) and wash length (L_b) where

43

berm aggradation begins is, $A_d = 0.138 \ L_b^2$. The constant, 0.138, is dimensionless.

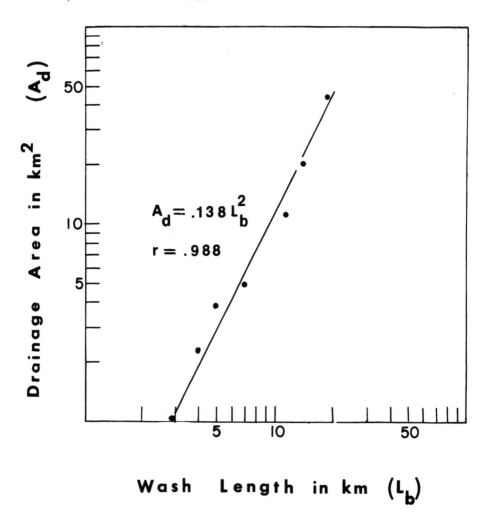

Figure 8. Logarithmic plots and regression analysis results for the drainage area and wash length above the apices of fine grained alluvial fans, or berm aggradation.

Sediment Source For Berm Aggradation

The grain size distributions on weathered alluvial fan interfluves of the upper piedmont shows that fine sediment (-1 to 3 ϕ) have been removed by either sheeterosion, eolian activity, or a combination of both. Additionally, field observations indicate a paucity of fine sediments in

44

the upper few centimeters of the surficial deposits (i.e. desert pavement) on the upper piedmont (Figure 1). The size of the sediment removed from the weathered alluvial fans on the upper piedmont is similar to the grain size distribution of berm sediments (1 to 4ϕ). It is inferred that the interfluves of the dissected, upper piedmont are the sediment source for the fine grained alluvial fans.

The amount of fine sediment yielded from one of these subwatersheds should be proportional to the amount of source area or interfluve widths. Those basins which have high values of mean interfluve width (W_I > 0.14 km) display fine grained fan aggradation, whereas, those subwatersheds with small values of mean interfluve width (W_I < 0.14 km) do not display berm aggradation (Table 1).

Additionally, the mean interfluve width can be expressed in terms of the drainage network where W_I equals the inverse of drainage density (D^{-1}). An increase in drainage density, or decrease in D^{-1}, reduces the source area for fine berm sediments and provides more channelways for efficient removal of these sediments from the subwatersheds.

Regional Implications of Berm Aggradation

The subwatersheds taken together provide a basis for drawing conclusions about geomorphic processes over the bolson plain of the entire Harquahala Valley. The areal distribution of the aggradational processes and their relationship to major physiographic features of the Harquahala Valley are described below.

Fine grained alluvial fans are interpreted as coalescing berm sediments which have spread laterally over the shallow interfluves of the lower piedmont. These fine grained fans represent areas of net aggradation in the Harquahala Valley and cover approximately 16 percent of the basin area (250 km^2) (Figure 1). The contact between the fine grained alluvial fans in the lower piedmont and coarse grained alluvial fans on the upper piedmont is transitional (Figure 1), but in general, the contact repre-

45

sents the lateral zero edge of alluviation on the bolson plain.

The sediment source for the fine grained fans is the fine sediment derived by sheetwash erosion or eolian processes from the coarse grained alluvial fans of the upper alluvial apron region, and to a lesser extent, the colluvium-mantled bedrock slopes of the mountain ranges (Table 1). The position of the zero edge of alluviation in the study area is, in part, controlled by the amount of sediment swept from the source area. The present ratio of aggrading to eroding bolson plain is 1:5, and the ratio of aggrading areas to the total source area (including the bedrock) is 1:8.

Fine grained fans are not evenly distributed over the center of the bolson plain (Figure 1). For example, the fine grained alluvial fans of the piedmont flanking the Eagletail Mountains have small areal extents as compared to those of the Big Horn Mountains, which have the most extensive areas of aggradation (Figure 1). Values for the mean interfluve width of watershed on the piedmont flanking the Eagletail Mountains are less than 0.140 km. This means less sheeterosion on interfluves and more channelized flow, resulting in more efficient removal of sediment from the piedmont. The Big Horn Mountains' piedmont has values of W_I greater than 0.140 km. Major physiographic differences between these areas are, 1) the Big Horn Mountains have larger bedrock source areas and a correspondingly higher piedmont length than the Eagletail Mountains (Wells, 1976), and 2) Centennial Wash is closer to the Eagletail Mountains than the Big Horn Mountains. The lithology of the source areas for both mountain ranges is similar (Table 1). Both differences are related to the history of the bolson plain, but in general, the Big Horn Mountains have a higher sediment yield. This probably diverted the axial drainage towards the Eagletail Mountains, which have had a lower sediment yield. In addition to influencing the relationship between drainage area and wash length, the impingement of Centennial Wash on the Eagletail Mountains' piedmont has increased the drainage density by causing accelerated stream erosion. Both factors reduce the condition which promotes berm aggradation.

SUMMARY AND CONCLUSIONS

Drainage area and wash length from the headwaters affect the location of deposition both in channel sedimentation and berm aggradation. These two geomorphic variables influence the transport distance for both coarse and fine sediments downwash. The source of berm sediments is eroded interfluves on the upper piedmont and the amount of fine sediment produced from these watersheds is proportional to the mean interfluve width. Thus, an increase of mean interfluve width, or decrease in drainage density, promotes an increase in fine sediment for fine grained fan deposition. The contact between the fine grained alluvial fans of the lower piedmont and the dissected, coarse grained alluvial fan of the upper piedmont represents the lateral zero edge of alluviation in the desert basin.

All ephemeral washes and their subwatersheds which display channel deposition at the local maximum width and berm aggradation in the Harquahala Valley follow these power functions:

Channel Deposition: $A_d = 2.5 \, L_c^{1.1}$

Berm Aggradation: $A_d = 0.138 \, L_b^{2}$.

Thus, these mathematical expressions may be used to predict areas of net deposition on alluvial fans of southwestern Arizona.

ACKNOWLEDGEMENTS

The manuscript has benefited from the careful reviews of Laurence Lattman of the University of Utah and William Bull of the University of Arizona. Robert Fleming of the U. S. Geological Survey, Paul Potter and Louis Laushey of

the University of Cincinnati were most helpful in sharing
their comments and ideas related to this study.

Financial support was provided by Sigma Xi, the Re-
search Society of North America; the Ohio Academy of
Science; the Summer Research Fellowship, University of
Cincinnati Research Council; and the Fenneman Fellowship,
Department of Geology, University of Cincinnati.

REFERENCES CITED

Babcock, H. M. and Cushing, E. M., 1942, Recharge to ground
    water from floods in a typical desert wash, Pinal
    County, Arizona: Am. Geophys. Union Trans. v. 23,
    p. 49-56.

Blissenbach, E., 1954, Geology of alluvial fans in semi-arid
    regions: Geol. Soc. America Bull., v. 65, p. 175-190.

Bull, W. B., 1964, Geomorphology of segmented alluvial fans
    in Fresno County, California: U. S. Geol. Survey Prof.
    Paper 352-E, p. 89-129.

_____, 1968, Alluvial fans, Jour. Geol. Ed., v. 16,
    p. 101-106.

Burkham, D. E., 1970, Depletion of stream flow by infiltra-
    tion in the main channelways of the Tucson Basin,
    southeastern Arizona: U. S. Geol. Survey Wat. Sup.
    Paper 1939B, 36 p.

Chow, V. T. (ed), 1964, Handbook of applied hydrology:
    McGraw-Hill Book Co., New York, 1418 p.

Denny, C. S., 1965, Alluvial fans in the Death Valley Re-
    gion, California and Nevada: U. S. Geol. Survey Prof.
    Paper 466, 62 p.

Debief, J., 1953, Les vents de sable dans le Sahara francais,
    in Actions Eoliennes, Cent. Nat. de Rech. Sci. Paris,
    Coll. Int., 35, p. 45-70.

Dunbier, R., 1968, The Sonoran Desert: University of Ari-
    zona Press, Tucson, 426 p.

Hack, J. T., 1957, Studies of longitudinal stream profiles
    in Virginia and Maryland: U. S. Geol. Survey Prof.
    Paper 294-B, 53 p.

Joly, F., 1953, Quelques phenomems d'ecoulement sur la
    bordure du Sahara dans les Confins Algero-Marocains
    et leurs Consequences Morphologiques, Compt. Rend.
    XIX Congres Geol. Intern., Algiers 1952, pt. VII,
    Deserts actuels et anciens, p. 135-146.

Leopold, L. B. and Miller, J. P., 1956, Ephemeral streams-
    hydraulic factors and their relation to the drainage
    net: U. S. Geol. Survey Prof. Paper 282-A, 37 p.

_____, Wolman, M. G., and Miller, J. P., 1964, Fluvial
    processes in geomorphology: W. H. Freeman & Co., San
    Francisco, 522 p.

Lustig, L. K., 1965, Clastic sedimentation in Deep Springs Valley, California: U. S. Geol. Survey Prof. Paper 352-F, p. 131-192.

McGee, W. J., 1897, Sheetflood erosion. Geol. Soc. America Bull., v. 8, p. 87-112.

Melton, M. A., 1965, The geomorphic and paleoclimatic significance of alluvial deposits in southern Arizona: Jour. Geol., v. 73, p. 1-38.

Renard, K. G. and Keppel, R. V., 1966, Hydrographs of ephemeral streams in the Southwest: Am. Soc. Civil Eng., no. HY2, Proc. Paper 4710, p. 33-52.

Ross, C. P., 1923, The Lower Gila Region, Arizona: U. S. Geol. Survey, Wat. Sup. Paper 498, 237 p.

Schick, A. P., 1970, Desert floods: interim results of observations in the Nahal Yael Research Watershed, southern Israel, 1965-1970: ISAH-UNESCO Sym. Res. on Rep. and Exp. Basins, New Zealand, p. 478-93.

Schumm, S. A. and Lichty, R. W., 1965, Time, space, and causality in geomorphology: Am. Jour. Science, v. 263, p. 110-119.

Siegel, Sidney, 1956, Nonparametric Statistics for the Behavioral Sciences: McGraw-Hill Book Co., New York, 312 p.

U. S. Geological Survey, 1970, Surface Water Supply of the U. S., 1961-1965: Wat. Sup. Paper 1926, Colorado River Basin, 571 p.

_____, 1975 Surface Water Supply of the U. S., 1966-1970, Wat. Sup. Paper 1926, pt. 9, Colorado River Basin, p. 681.

Wells, S. G., 1976, A study of surficial processes and geomorphic history of a basin in the Sonoran Desert, Southwestern Arizona: unpbl. Ph.D. Dissertation, University of Cincinnati, 328 p.

Werho, L. L., 1967, Compilation of flood data for Maricopa County, Arizona, through September 1965: Maricopa County, Ariz., 36 p.

THE FORMATION OF PEDIMENTS: SCARP BACKWEARING OR SURFACE DOWNWASTING?

John H. Moss

Department of Geology
Franklin and Marshall College

ABSTRACT

The study of the formation of pediments in the south-western United States has a long evolution. Early geomorphologists envisaged a general model in which the dominant process was backwearing of a mountain scarp accompanied by concurrent development of an erosion surface (pediment) extending outward from its base. Pediments were believed to be surfaces of transport across which detritus produced in the mountains was carried to adjacent basins. Recent work has brought out that pediments vary in character and are formed in a variety of ways over different lengths of time. In south-central Arizona near Phoenix, some granitic pediments are composed principally of residual grus more than 60 feet thick derived from underlying granitic bedrock, whose irregular surface has little relation to erosion by lateral planation, rill action or sheetwash. The model outlined here emphasizes the importance of downwasting by weathering to explain the origin of these granitic pediments. Vertically the pediments can be differentiated into 2 distinct parts; a residual grus upper zone of variable thickness providing the relatively level surface topography of the pediment and a fresh bedrock zone below containing relief either related to the shape of the original granite mass or produced by differential weathering. As the original granite surface decomposed, probably during times of a wetter climate, the increasing thickness of grus accumulated. Mass wastage and minor movements by water caused the grus to fill in the original irregularities of the granite surface. The irregular bedrock surface below the grus is the uneven downward moving front of a weathering zone.

51

Residual hills, although in some cases completely "grusi-
fied" within, stand up above the pediment due to a protec-
tive armor of boulders. Where fresh bedrock is exposed,
particularly near the mountains, the grus has been removed
and present-day processes are modifying the surface of the
pediments.

INTRODUCTION

In 1912 a noted geologist, Sidney Paige, commenting
on Davis' "Geographical Cycle in an Arid Climate" (Davis,
1905), stated enthusiastically that the system erected by
Davis was so complete in its larger features that those who
followed needed only to assign particular physiographic
products to their proper place in an already established
larger system (Paige, 1912). In the ensuing 65 years it
has become evident that desert geomorphology is far more
complex than originally perceived. A growing body of fac-
tual data has been gathered which has either refined or cast
doubt on earlier theories of landform development based on
casual observations and inferences (Cooke and Warren,
1973). With respect to pediments in the Phoenix area, new
data raise questions about earlier theories which explained
pediment formation entirely in terms of infrequent fluvial
processes operative in the desert today. My own observations have led to the conclusion that the role of
weathering has been greatly underestimated and that at
least part of the history of pediment formation took place
in the distant past, possibly under different climatic
conditions than the present. Recent processes may merely
be modifying landforms produced in the past.

It is the purpose of this paper to provide a summary,
albeit incomplete, of the status of work on the origin of
pediments in the southwestern United States and to add
some new data from the Phoenix area in south-central
Arizona.

Geographic Setting

The Phoenix area lies in the Basin and Range physiographic province on the northeastern margin of the Sonoran
desert (Figure 1). This province consists largely of a
series of linear mountain ranges of diverse lithology separated by sediment-filled basins. Rainfall varies with altitude but is generally less than 10 inches annually at lower
elevations, with much of the rainfall coming in cloudbursts
widely separated in time and space. Of particular interest
in this paper is the piedmont zone where the mountains meet
the basins. Here is found one of the most enigmatic of all
desert landforms, the pediment, a gently sloping rock surface whose origin has been the subject of a multitude of
studies and about which there is still not complete agreement. According to Cooke (1970), 30 percent of western
Arizona deserts is composed of exposed pediment and according to Corbel (1963) 10 percent more is "covered pediment."
In addition to pediments, the other major piedmont landforms
are alluvial fans occurring at the mouths of ravines and
canyons where steep-gradient mountain streams spread their
clastic load on the adjacent basin floor in a fan-shaped
deposit. In areas where bedrock is not exposed, Denny
(1967) has emphasized the difficulty of deducing from the
surface whether the landform is a fan or pediment. Doehring
(1970) has demonstrated that through analysis of the shape
of contour lines on topographic maps this distinction can
be made in many cases.

Unlike pediments developed in hard rock, which are
part of an extremely slowly evolving system, many fans are
actively forming today, and a process-form system is more
readily detectable (Denny, 1967; Bull, 1963, 1968). Their
origin is the result of changes in the hydraulic geometry
of flow after the stream leaves its channel at the mountain
front and spreads out on the growing fan. Decrease in
depth and velocity of flow are critical variable parameters.
In contrast, the origin of pediments is still far more conjectural.

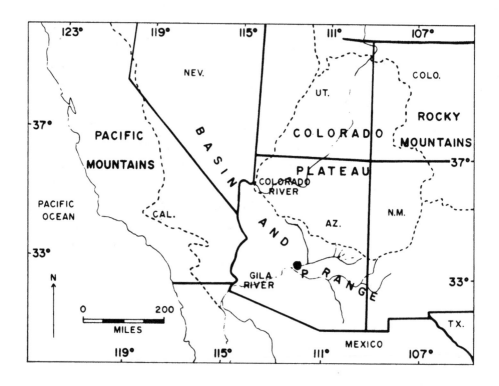

Figure 1. The Southwestern United States. The large dot and letter "P" show the location of Phoenix, Arizona. It is situated near the northeast corner of the Sonoran Desert.

The Pediment Problem

The pediment problem does not deal with the origin of an alluvial feature but with the origin of a gently sloping (commonly less than 5°) bedrock surface developed on resistant rock in an arid region. Pediments commonly terminate against a more steeply rising (greater than 20°) mountain front. Since both the mountains and pediments are underlain by the same resistant rock, the problem of pediment origin centers on how this surface of low relief was developed from the mountain massif against which it abuts.

Over the years, because of the inability of geomorphologists to view the evolution of large-scale pediments developed in hard rocks, it has been difficult to know which of their characteristics are definitive in determining their origin. Because of this dilemma it has been possible by selective weighting of observations or measurements to come up with many conflicting hypotheses for their evolution. (For summaries, see Tator, 1952; Tuan, 1959; Cooke and Warren, 1973).

Three shortcomings emerge from the plethora of published pediment studies. First, there is no longer general agreement on the meaning of the term "pediment." Either the definition has been broadened to include almost any erosion surface anywhere, or pediments are defined in terms of a set of processes believed to have formed them. Attempts to improve the precision of pediment terminology such as Howard's introduction of the terms "peripediment" and "pediplane" in his study of the Sacaton Mountains of Arizona (Howard, 1942), although helpful, have not come into common usage.

Secondly, there seems to have been a tendency to champion one process, or set of processes, to explain the origin of all pediments. There is an implication of universality in the mode of pediment formation proposed. That pediments might be diverse complex landforms of varied origin and history appears to have been overlooked.

Thirdly, until recently, serious attention has not been given to the idea that many pediments may be relict

features. Their history may have included a long period of
weathering granite to grus and the "erosion", "planation"
or "truncation" freely claimed by virtually all writers to
have produced the bedrock pediment surface may instead have
principally been directed toward removing a thick cover of
grus or soil. It is with the last idea that this paper will
be particularly concerned.

Characteristics of Pediments

For the purpose of this paper, a slightly modified ver-
sion of Hadley's definition of a pediment will be used.
"Pediments are bedrock surfaces of low relief, partly
covered by a veneer of rock waste, that slope from the
base of mountain masses or escarpments in arid or semi-arid
regions" (Hadley, 1967). Admitting considerable variation
from the norm, in its simplest form, a pediment is a rock
bench of varying width sloping at a low angle (usually less
than 5°) from the foot of a mountain front into the adjacent basin. Growing tongue-like extensions of the surrounding pediment may reach into a mountain range from all
sides, in time cutting it up into a series of residual
hills.

At some localities the angle of intersection of pediment and mountain front is greater than 30°, whereas elsewhere pediments join adjacent mountains in smooth curves.
The mountainward edge of a pediment is commonly a rock
apron covered by only patches of rock waste. Basinward,
rock waste becomes more widespread and thicker, finally
mantling the granite so deeply that it becomes a "suballu-
vial bench" (Lawson, 1915). From this point outward it is
impossible to tell without geophysical help whether one is
on pediment or the main alluvial fill of the basin beyond
the pediment margin.

Both near and far from the mountain front on the Usery
Mountain-Pass Mountain Pediment east of Phoenix, granite
hills rise 40 to 300 feet above the thinly veneered pediment
surface, a frequent occurrence elsewhere in the area.

Faulting is common in the Basin and Range Province and has been used to explain the original scarps whose retreat is believed to have resulted in pediment development. Seldom, however, in arroyos or on the pediments or around residual hills is it possible to locate the border faults with respect to mountain or pediment.

Pediments are not planar surfaces but are either slightly etched by a network of anastamosing channels a few inches to a foot deep, or cut by deeper channels with tributaries cutting dendritic drainage pattens into the upper edge of the pediment (Gilluly, 1937).

The pediments near Phoenix are developed in the same resistant rocks that comprise the mountains with which they merge. Longitudinally, pediments are convex upward, increasing in steepness toward the mountains. Parallel to the mountain fronts, Gilluly (1937) has shown that the pediments of the Ajo Mountains of Arizona are either concave or virtually planar. Only in passes where pediments from opposite sides of a mountain range join are they commonly convex.

FORMATION OF PEDIMENTS

Much of the pediment research in the last 100 years in the southwestern United States has been centered on two problems: how did the bedrock surface of the pediment form, and how did the mountain front retreat while maintaining a sharp angle of contact with the pediment? Early workers (Gilbert, 1877; McGee, 1897 and Lawson, 1915) profoundly influenced later thinking by proposing the preeminence of backwearing of a mountain scarp parallel to itself, leaving a relatively level pediment stretching basinward from its foot. As the mountain slopes continue to retreat and the pediment grows mountainward, its basinward margin becomes increasingly covered by alluvium. The pediment seems to be almost universally thought of as a cut surface, the principal difference in proposed origins lying in the cutting process or processes envisaged. Gilbert (1877),

57

who first described thinly veneered rock surfaces surrounding desert mountains in the Southwest, ascribed their
origin to lateral planation. McGee (1897) attributed their
cutting to sheetflooding after witnessing a torrential
flood coursing down a pediment in Arizona. Davis (1938)
proposed stream flooding in addition to sheetflooding as
important erosional agents in pedimentation. Rich (1935)
added "unconcentrated sheetwash, as contrasted with defin-
ite streams" as playing a leading part in erosion. He considered a sheetflood as an exaggerated form of sheetwash.
Bryan (1923) emphasized the importance of rill cutting in
moving debris down the mountain front and from its base
across the pediment toward larger streams. He also recognized the importance of lateral corrasion by larger ephemeral streams in pediment formation. Blackwelder (1931),
Johnson (1932), Field (1935), Gilluly (1937), Davis (1938),
Sharp (1940), Howard (1942) and Rahn (1967) all proposed
pediments to be planation surfaces with lateral corrasion
playing a primary role. Gilluly (1937) raises the question
that since tributary streams and rill action are the processes chiefly lowering the interstream areas today, may
the pediments not have been "born dissected"?

It is interesting to note that although erosive agents
truncating the resistant rocks of the pediment are explained in considerable detail by most investigators,
little, if any mention is made of what happens on the pediment between these infrequent events. Despite the ubiquitous occurrence of grus, weathering is seldom described '
or stressed as an important process. Pediments are viewed
as surfaces of transportation, not as a substantial source
of sediment themselves. Only King (1953) raises the question of whether these truncating processes are capable of
explaining the formation of pediments and the retreat of
their "commanding hillsides." He proposes that they become effective only after the pediment is in existence.
Deep weathering of pediment surfaces probably contributed
to their downwearing independent of scarp backwearing.

Attempts have been made to explain the difference in slope between mountain front and pediment on the basis of the size of rock waste mantling each. However, despite positive conclusions by Lawson (1915), Bryan (1923), Gilluly (1937) and Rahn (1966), quantitative studies by Mammerickx (1964), Cooke (1970) and Cooke and Reeves (1972) in the Mojave Desert and Melton (1957) in southern Arizona failed to find a positive correlation between slope angle and surface rock size. In fact, most morphometric studies of parameters of the pediment-mountain system, for example, slope and pediment length, slope angle and size of drainage basin backing the pediments show a poor relationship between present-day form and process. This raises further questions as to whether present-day processes alone are accountable for pediment development.

In addition to uncertainties about the origin of pediments are doubts about the mode of retreat of mountain scarps and the fate of the angle of junction between pediment and mountain in the southwestern U. S. Lawson (1915) proposed the retreat of the scarp, or frontal slopes, of the mountain parallel to itself in accord with its lithology and the weathering processes acting on it. In the Papago country of western Arizona, Bryan related the scarp slope angle to the resistance of the rocks composing the mountain front and the spacing of joints. He noted that boulder controlled slopes in massive granite, gneiss and conglomerate had declivities ranging from 20°-45°. He found mountain slopes less than 20° to be rare and related to closely jointed gneiss, schists, phyllite and felsite. The lower declivity of the pediment is related to the smaller caliber of the sediments covering it (Cooke and Reeves, 1972).

Johnson (1932) postulated extensive lateral cutting to produce retreat of the mountain front concurrently with the development of the pediment, but visualized a decline in the angle of its slope with time. However, sharp piedmont angles occur along segments of the mountain front without streams. The reliance by King (1953) on the contrast be-

59

tween turbulent flow on scarps and laminar flow on pediments is not supported by observations of flow regime on
pediments after summer cloudbursts. Lustig (1969) associated scarp retreat with its steep slope which contrasts
with the lower gradients of the master channels of pediments extending into the mountains. Being a steep inter-
fluvial area subject to weathering, scarps bordering pediment extensions into the mountains are worn back more
rapidly because the fewer weathering products on the steep
slopes could be more readily removed by short duration runoff.

Schumm's measurement of pediment erosion and slope retreat on miniature pediments in Badlands National Monument
over a period of 8 years showed that the pediments were
lowered by sheetwash and the scarp retreated, leaving a
belt of newly formed pediment 8-12 cm wide (Schumm, 1962).
This study confirmed the earlier work of Smith (1958) on
the importance of slopewash in pediment formation and
scarp retreat in these formations. However, whether measurements of miniature pediment formation in the silts and
clays of the Chadron and Brule formations are applicable
to the large-scale granitic pediments and scarps of Arizona
is questionable.

Twidale (1967) has pointed out correctly from observations in south Australia that the backwearing of the escarpment and the resultant development of a flat pediment
is highly complex involving many factors. He has also
noted that well developed angles form between scarps and
pediments in areas where no faulting has occurred. In the
Basin and Range Province of the United States faulting is
prevalent but seldom corresponds with present-day scarps
bordered by pediments. Twidale documents cases where once
formed, a hill plain junction becomes a zone of moisture
concentration. The scarp-foot zone is gradually worn down,
the scarp steepens and slowly retreats.

Regarding the rate of scarp retreat, relations at
the sites of radiometrically dated woodrat middens in

60

Joshua Tree National Monument reported by Oberlander (1974) indicate no appreciable scarp retreat in quartz monzonite in the last 10,000 years. In the Alabama Hills of Owens Valley, granitic outcrops show no evidence of backwearing in approximately 50,000 years, since the Tahoe age glacial outwash accumulated against them (Oberlander, 1972). The rate of removal of quartz monzonite beneath radiometrically dated basalt in the adjacent White Mountains indicates rates of erosion ranging from 10 to 30 mm/1000 years over the past 10.8 m.y. (Marchand, 1971). Removal of well-decayed granodiorite and quartz monzonite from beneath basalt dated at 3.9 ± 0.9 m.y. in the Lucerne Valley of the western Mojave Desert revealed a maximum erosion rate of approximately 8 mm/1000 years. At these rates, scarp retreat of over 2 miles in Arizona indicates tremendous antiquity for the pediments, if indeed they were formed primarily by scarp backwearing.

The Usery Mountain-Pass Mountain Pediment

Twenty-five miles east of Phoenix a prominent pediment borders the Usery Mountains and Pass Mountain (Figures 2 and 3). The latter mountain is a prominent landmark because of a thick layer of rhyolite clearly visible on its upper slope. These mountains reach elevations slightly above 3200 feet.

The pediment, rising from approximately 1600 to 2100 feet on the south side of the mountain, is developed on the same resistant granitic rocks that form the bulk of the mountains. Granite also comprises the 15 or more boulder-covered residual hills standing 40 to 300 feet above the pediment at distances up to 3 miles from the mountain fronts (see U.S.G.S. Buckhorn and Apache Junction, $7\frac{1}{2}$", Quadrangles). A tongue of the pediment extends upward from the broad basin to the south forming Usery Pass (Figure 2). Here it merges with a pediment rising from the Salt River Valley to the north described by Péwé (1970).

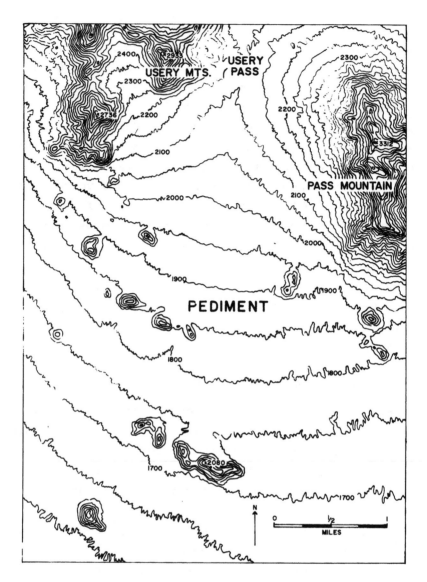

Figure 2. *Topographic map of pediment with residual hills 25 miles east of Phoenix, Arizona bordering the Usery Mountains and Pass Mountain on the south. Contour interval is 100 feet.*

Three miles from the mountain front the pediment is smooth and grus-covered, sloping only 50 feet per mile, $\frac{1}{2}°$ (Figure 4). Closer to the mountain its surface becomes rougher, more bedrock is exposed at the surface, the grus cover becomes patchy and its slope increases to 500 feet per mile, 3½° (Figure 5). Continuing upward, the slope rises rapidly attaining 20° to 30° at the mountain fronts, which are notably jointed with cliffs and large blocks abounding.

Figure 3. View eastward of Pass Mountain Pediment.

A few prominent arroyos about 1/3 mile apart form steep-sided channelways averaging about 10 feet deep in the central part of the pediment. Less well-defined waterways used infrequently by the scattered rains follow anastomosing courses on the grus-covered interfluves between the arroyos.

Of greatest significance to determining the formation of the pediment is the interpretation of the grus and the irregularity of the bedrock surface beneath it. The grus is sand to small pebble size consisting of individual mineral grains or clusters of grains derived from the underlying granite.

Wahrhaftig (1965) attributes the formation of grus on the western flank of the Sierra Nevada to swelling, accompanying decomposition of biotite and hydration of feldspar which is powerful enough to disintegrate the granite. Studies by Eggler, Larson and Bradley (1969)

63

Figure 4. *Smooth grus-covered pediment surface about 2½ miles from Pass Mountain front.*

Figure 5. *Rough dissected section of pediment about 3/4 mile from Pass Mountain front. Flatter areas of bedrock are thinly veneered with grus. Some bedrock surfaces are friable having undergone partial decomposition.*

on the production of grus in the Trail Creek granite of the Sherman Peneplain in the Laramie Range, Colorado-Wyoming, connect its formation with the expansion of biotite as it alters to hematite between cleavage planes and along contacts with other minerals.

Three miles south of Usery Pass is an irregularly bottomed bedrock pit from which the overlying grus has been excavated by power shovel (Figures 6 and 7). Note the deceptively flat surface of the pediment north of the pit. The depression contours outline the uneven configuration of the bedrock topography beneath the grus, which varies in thickness from 0-12 feet at this locality. Mineralogically, the grus resembles the underlying bedrock. Occasional small unweathered and partially weathered blocks litter the slopes and floor of the pit.

The irregularity of the bedrock surface is probably due to slight differences in the mineral composition of the rock and the length of time moisture remained in contact with the bedrock. It is a "weathering front" beneath the grus.

The relief of the fresh bedrock surface is also represented by the bedrock which probably, but not positively, forms the cores of the boulder-clad residual hills rising above the pediment (Figures 2 and 8) and by additional irregularities in the granite surface visible in arroyo walls.

Much of the grus appears to be residual, having weathered from the bedrock directly below it. It is commonly unstratified without sedimentary flow structures. Some of the upper part, however, may have washed downslope during infrequent rains. The granite bedrock is thus the C horizon of a soil profile and the grus the lower part of an overlying horizon. The remainder of the profile, which may have had considerable thickness, has been gradually washed toward the center of the basin.

What we see today is the remnant of a subsurface soil profile. Twidale (1962) has described weathering of granite in arid, south Australia as "soil moisture weathering."

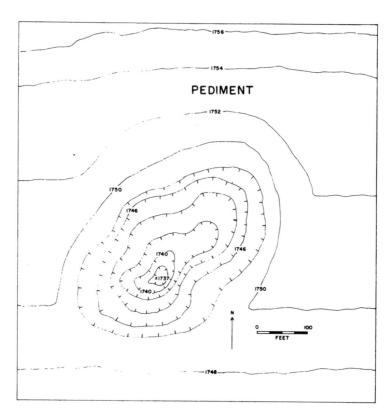

Figure 6.   *Map of bedrock pit from which grus has been removed for commercial purposes by power shovel.   Contour interval is 2 ft. Note difference between smooth pediment surface and irregular configuration of bedrock, which hardly resembles topography cut by running water.*

Figure 7.   *Surface view of pit in Figure 6 showing irregular bedrock topography of excavated area.*

Although Phoenix is an area of low rainfall and high summertime temperatures, dew in the winter and water soaking into the regolith during the occasional summer rainstorms have enough longevity to accomplish the chemical weathering which disintegrates the granite.

In the wetter climates of the past in the Southwest, decomposition may have been accelerated and much larger quantities of grus formed as long as the upper layers of the soil zone were not plugged with caliche, preventing downward percolation of surface water. Because grus is easily transported by running water, sheetfloods or laterally cutting streams might remove it. On the other hand, it seems unlikely that such ephemeral streams could significantly corrade granite or produce the unusual irregular bedrock topography of the granite surface.

That life zones and hence moisture zones have shifted vertically in the Southwest in the past is attested by marmot finds 2000 to 3000 feet below their present range in New Mexico (Stearns, 1942) and Pleistocene woodrat middens in low ranges in the Mojave Desert of southern Nevada which contain leafy twigs and juniper berries, indicating a significant climate change (Wells and Jorgenson, 1964). Three pollen spectra from lake sediments below mid-Kansan vertebrate fauna near Safford, Arizona indicate plant associations approximating those found 1500 feet above the desert floor of the Sonoran desert today (Gray, 1961).

The present thickness of grus on this pediment (even thicker deposits probably existed in the past) together with the degree of bedrock surface relief, point to two conclusions about the pediment's origin. First, weathering present and past has played a major role in the formation of the pediment. Secondly, running water, although important in cutting deeper channels in the pediment and to some degree eroding and washing the grus downslope, could hardly have been a dominant laterally corrading agent in forming the uneven bedrock surface below the grus. Pedi-

ment stream courses are relatively straight, lacking the
channel pattern for lateral cutting.

*Figure 8. Boulder-covered residual hill rising above Usery Mountains-
Pass Mountain Pediment. Boulders are formed by weathering
of jointed granite bedrock comprising hill. Smooth surface
in foreground is grus-covered slope. Grus consists of disintegrated granite particles washed from hill and weathered
from underlying pediment.*

In contrast to the retreating scarp model for pediment
formation proposed by Lawson (1915) and later geologists
(Figure 9A), the model proposed here involves a two-stage
downwearing concept including an initial period of deep
weathering followed by erosion and continued but less intense weathering to form the present topography (Figure 9B).
As depicted, the present topography is believed to reflect
to some extent the original topography of the granite mass
indicated by the dotted line.

Oberlander (1974) has presented evidence for a somewhat similar two-stage mechanism for pediment development
in the Mojave Desert. Evidence indicates that the pediments
there are relict features inherited from a wetter, semi-arid
climate in the Tertiary when granitic slopes were covered
by saprolite which has been stripped off by Quaternary erosion. The preservation of Tertiary weathering profiles

preserved under lava flows dated at 8 m. y. ago indicate that the pediments had been formed at that time. In southeastern Arizona, Tuan (1962) believes present-day pediments to have been formed by stripping of overlying materials.

Determination of the age of pediment formation in the Pass Mountain area awaits more detailed checking of the condition of the granite beneath the Tertiary volcanics capping Pass Mountain which have been dated at 22 m. y. Preliminary reconnaissance indicates deep weathering of the granite which, if correlative with the weathering products produced on the pediment, indicates a mid-Tertiary age for the original pediment. Later Tertiary and Quaternary denudation has removed much of the regolith, leaving the bedrock exposed on the pediment and mountain. According to this model, the role of Quaternary weathering and erosion has been to accentuate the slope difference between mountain and pediment.

As concluded by Mabbott (1966) in central Australia, sheetflow appears incapable of producing erosional planation of fresh granite but is capable of smoothing irregularities in the surface by small movements and deposition of rock waste. As stated above, the relief of the pediment bedrock surface argues against its being formed by lateral planation but not against its being the uneven front of a downward moving weathering zone (the grus) which has been totally removed in the areas where bedrock is exposed.

It is not possible to tell how much of the unevenness of the bedrock is due to differential weathering, and erosion, and how much is inherited from the original topography of the granitic body. As downwasting of the granite progressed, as shown in Figure 9B, the grus may have been considerably thicker than it is now. A shift to the present drier climate has increased runoff and enable ephemeral streams, sheetfloods and sheetwash gradually to remove most of it from both this pediment and mountain front, leaving the present pediment-mountain front relationship. The present-day pediment is largely a surface of transport for

69

material weathered from it. Finer material from the mountains moves principally down the gullies and arroyos.

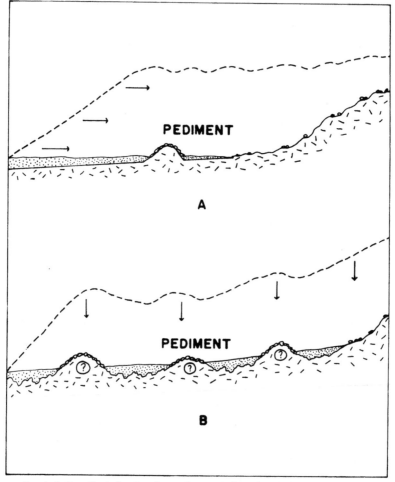

Figure 9. A & B Sketch showing alternative hypotheses of pediment
    formation. "A" portrays formation by backwearing
    of original scarp (arrows) leaving pediment at its
    foot. "B" stresses a period of downweathering (arrows)
    followed by erosion to form present irregular bed-
    rock topography, boulder-clad original hills, and
    remnants of originally thicker grus.

Union Hills Sand Pits

    Fifteen miles north of Phoenix, east of the Black Can-
yon Highway, in the pediments leading eastward to the Union
Hills, are sand pits from which large quantities of grus
have been removed for commercial purposes. One of these

70

operations which has cut through an unnamed residual knob is of particular interest in determining the geologic history of the area.

As shown in Figures 10 and 11, a pit 37 feet deep has been excavated by power shovels into the friable grus overlying granite bedrock. A drill hole northeast of the pit indicated a total grus thickness of 60 feet before bottoming in granite.

Figure 10.   *Grus pit 37 feet deep in granite pediment near Union Hills north of Phoenix. At this locality, grus is 60 feet deep.*

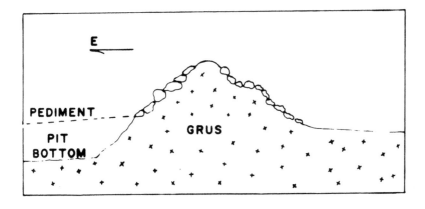

Figure 11.   *Diagrammatic cross-section showing pit and adjacent hill which is mantled with boulders but has a core of grus indicating a total depth of weathering of more than 100 ft.*

More surprising than the thickness of grus in the pediment is the weathered condition of the 90-foot high residual hill, the north and east sides of which are being excavated. Although mantled by only slightly weathered granite boulders, the entire inside of the hill consists of loosely compacted grus easily removed by power shovels (Figure 12). The boulders exposed on the surface protect the friable grus from erosion by the occasional heavy desert rainstorms. They also protect from evaporation, the moisture which percolates into the porous grus and further weathers the underlying granitic waste.

Figure 12. Excavation into hillside shown in Figure 11. Note boulders on surface overlying deeply decayed granite bedrock converted mostly to friable grus.

The occurrence of deeply weathered granite below granite boulders resembles similar field relationships in the stepped topography of the Great Valley and Sierra Nevada in California described by Wahrhaftig (1965). He noted that buried granitic rocks weathered more rapidly than unburied because exposed rock is dry most of the year; whereas, buried granite is kept moist with reactive solutions proceeding inward from joint surfaces grain by grain, producing disintegration to grus. Although the sand size grus

is readily transported by small water flows, the unweathered joint blocks form a lag on the surface. This
weathering process produces the bimodel sediment size distribution so common in granite terrains. Oberlander (1972)
has described similar occurrences of advanced chemical alteration of subsurface granitic materials, some scores of
feet deep, below boulder mantles on residual hills in the
Mojave Desert.

The great thickness of grus described here once again
indicates the importance of weathering and downwearing in
the formation of granite pediments in this area. Where the
grus is protected by boulders, it remains in hills and
ridges. Where the boulders have finally weathered to a
critical size of about 1 foot (Oberlander, 1974), they disintegrate and the grus is exposed to erosion and is washed
toward lower areas of the basin. In time, in this area
near the Union Hills, stripping of grus will proceed to a
level where the unweathered granite beneath the grus will
be exposed and the landscape will more nearly resemble the
Usery Mountains-Pass Mountain area.

The formation of such a great thickness of grus indicates a long period of weathering, probably some of it in
a wetter climate. The presence of the grus, which appears
to be a residual deposit on the pediment, apparently rules
out major action by running water in cutting the underlying bedrock. Backwearing of a scarp across this area,
leaving a bedrock pediment at its foot, is difficult to
visualize. Downweathering producing grus, which is later
removed, seems more likely.

CONCLUSIONS

1. Many earlier studies of granite pediments center on the
   present-day processes affecting their surfaces and the
   mountain fronts to which they abut. These works may
   have overlooked that present-day activity is merely the
   latest chapter in a long complex history producing
   these landforms. Furthermore, part of their history

73

included a time when the climate probably was less arid than at present.

2. Study of the granite pediment which slopes southward from the petrologically similar Usery Mountains and Pass Mountain east of Phoenix reveals that the smooth grus-covered pediment surface is underlain by a bedrock topography too irregular to have been formed by sheet-flow, rill action, or lateral planation.

3. Residual grus 60 feet thick comprising the upper part of a pediment near the Union Hills north of Phoenix and forming the core of a boulder-clad, partly excavated adjacent hill reveals total granite weathering in this locality of over 100 feet. This grus thickness does not fit with the conception of "a pediment as a surface of transportation."

4. These discoveries cast further doubt on the formation of these pediments by the backwearing of scarps with concurrent truncation of a bedrock platform at their foot by erosional agents.

5. The great depth of weathering represented by the thickness of decayed and disintegrating granite indicates considerable antiquity for the pediments. Such a thickness supports the idea that they were subjected to a long period of weathering, producing the grus, followed by a stage of stripping revealing their present granite surfaces. The irregular bedrock topography beneath the grus represents in some places an advancing downward-moving weathering front.

ACKNOWLEDGMENTS

The ideas presented in this paper were germinated by field observations while on tenure as a Visiting Professor at Arizona State University in 1975. I am grateful to many members of the staff of the Geology Department, particularly Michael Sheridan, Troy Péwé, Robert Lundin, Donal Ragan, Carleton Moore, Peter Buseck, and William Sauk, for discussions about the geology of the area and for familiarizing me with some of its intricacies. I take full responsibility, however, for the ideas presented here which in no sense necessarily represent those of the staff.

REFERENCES CITED

Blackwelder, E., 1931, Desert plains: Jour. Geol., v. 39, p. 133-140.

Bryan, K., 1923, Erosion and sedimentation in the Papago country: U. S. Geol. Survey Bull. 730, p. 19-90.

Bull, W. B., 1963, Alluvial fan deposits in western Fresno County, California: Jour. Geol., v. 71, p. 243-251.

_____, 1968, Alluvial fans: Jour. of Geol. Ed., v. 16, p. 101-106.

Cooke, R. U., 1970, Morphometric analysis of pediments and associated landforms in western Mojave desert, California: Amer. Jour. Science, v. 269, p. 26-38.

_____ and Reeves, R. W., 1972, Relations between debris size and slope of mountain slopes in the Mojave desert, California: Zeit. fur Geomorph., v. 16, p. 76-82.

_____ and Warren, A., 1973, Geomorphology in Deserts: Univ. California Press, Berkeley, 394 p.

Corbel, J., 1963, Pediments d'Arizona: Centre de Doc. Cart. et Geog., Memoires et Documents v. 9, p. 33-45.

Davis, W. M., 1905, The geographical cycle in an arid climate: Jour. Geol., v. 13, p. 381-407.

_____, 1938, Sheetfloods and streamfloods: Geol. Soc. America Bull., v. 49, p. 1337-1416.

Denny, C. S., 1967, Fans and pediments: Am. Jour. Science, v. 265, p. 81-105.

Doehring, D., 1970, Discrimination of pediments from alluvial fans: Geol. Soc. America Bull., v. 81, p. 3109-3116.

Eggler, D. H., Larson, F. E. and Bradley, W. C. 1969, Granite grusses and the Sherman erosion surface, southern Laramie Range, Colorado-Wyoming: Am. Jour. Science, v. 29, p. 313-322.

Gilbert, G. K., 1877, Report on the geology of the Henry Mountains: U. S. Geog. Geol. Survey of the Rocky Mountain region: p. 99-150.

Gilluly, James, 1937, Physiography of the Ajo region, Arizona: Geol. Sco. America Bull., v. 48, p. 323-348.

76

Gray, J., 1961, Early Pleistocene paleoclimate record from Sonora Desert: Science v. 133, p. 38-39.

Hadley, R. F., 1967, Pediments and pediment-forming processes: Jour. Geol. Ed., v. 15, p. 83-89.

Howard, A. D., 1942, Pediment passes and the pediment problem: Jour. Geomorph., v. 5, p. 3-31, 95-136.

Johnson, D. W., 1932, Rock planes in arid regions: Geog. Rev., v. 22, p. 656-665.

King, L. C., 1953, Canons of landscape evolution: Geol. Soc. Ameria Bull., v. 64, p. 721-752.

Lawson, A. C., 1915, The epigene profiles of the desert: Univ. Calif. Dept. Geology Bull., v. 9, p. 23-48.

Lustig, L. K., 1969, Trend surface analysis of the Basin and Range province and some geomorphic implications: U. S. Geol. Survey Prof. Paper 500-D.

Mabbutt, J. A., 1966, Mantle-controlled planation of pediments: Am. Jour. Science, v. 264, p. 78-91.

McGee, W. J., 1897, Sheetflood erosion: Geol. Soc. America Bull., v. 8, p. 87-112.

Marchand, D. E., 1971, Rates and modes of denudation, White Mountains, eastern California: Am. Jour. Science, v. 270, p. 109-134.

Mammerickx, J., 1964 Quantitative observations on pediments in the Mojave and Sonoran deserts: Am. Jour. Science, v. 262, p. 417-435.

Melton, M. A., 1957, Debris-covered hillslopes of the southern Arizona desert-consideration of their stability and sediment contribution: Jour. Geol., v. 73, p. 1-38.

Oberlander, T. M., 1972, Morphogenesis of granitic boulder slopes in the Mojave Desert, California: Jour. Geol., v. 80, p. 1-20.

_____, 1974, Landscape inheritance and the pediment problem in the Mojave Desert of southern California: Am. Jour. Science, v. 274, p. 849-875.

Paige, S., 1912, Rock-cut surfaces in desert ranges: Jour. Geol., v. 20, p. 442-450.

Pewe, T., 1971, Guidebook to the Salt River Valley: Department of Geology, Arizona State Univ., Tempe, Arizona.

Rahn, P. H., 1966, Inselbergs and knickpoints in southwestern Arizona: Zeitschr. fur Geomorph., v. 10, p. 217-255.

_____, 1967, Sheetfloods, streamfloods and the formation of pediments: Ann. Amer. Assoc. Geog., v. 57, p. 593-604.

Rich, J. L., 1935, Origin and evolution of rock fans and pediments: Geol. Soc. America Bull., v. 46, p. 999-1024.

Schumm, S. A., 1962, Erosion on miniature pediments in Badlands National Monument, South Dakota: Geol. Soc. America Bull., v. 73, p. 719-724.

Sharp, R. P., 1940, Geomorphology of the Ruby-East Humbolt Range, Nevada: Geol. Soc. America Bull., v. 51, p. 337-371.

Smith, K. G., 1958, Erosional processes and landforms in Badlands National Monument: Geol. Soc. America Bull., v. 69, p. 975-1008.

Stearns, C. E., 1942, A fossil marmot from New Mexico and its climatic significance: Am. Jour. Science, v. 240, p. 867-878.

Tator, B. A., 1952, Pediment characteristics and terminology: Ann. Am. Assoc. Geog., v. 42, p. 295-317.

Tuan, Y. F., 1959, Pediments in southeastern Arizona: Univ. Calif. Pub. Geography, v. 13, 163 p.

_____, 1962, Structure, climate and basin landforms in Arizona and New Mexico: Ann. Amer. Assoc. Geog., v. 52, p. 51-68.

Twidale, C. R., 1962, Steepened margins of inselbergs from northwestern Eyre Province, south Australia: Zeitschr. fur Geomorph., v. 6, p. 51-99.

_____, 1967, Origin of the piedmont angle as evidenced in south Australia: Jour. Geol., v. 75, p. 339-441.

Wahrhaftig, C., 1965, Stepped topography of the southern Sierra Nevada, California: Geol. Soc. America Bull., v. 76, p. 1165-1189.

Wells, P. V. and Jorgenson, C. D., 1964, Pleistocene wood rat middens and climatic change in Mojave Desert - a record of juniper woodland: Science v. 143, p. 1171-1174.

78

ORIGIN OF SEGMENTED CLIFFS IN MASSIVE SANDSTONES
OF SOUTHEASTERN UTAH

Theodore M. Oberlander
Department of Geography
University of California, Berkeley

ABSTRACT

The spectacular cliffs of the Plateau Country of southeastern Utah generally result from the collapse of massive
sandstone caprock due to loss of support following weath-
·ering and erosion of subjacent thin-bedded sandstones and
shales. However, prior to the exposure of a weaker substrate, major slope breaks develop within the massive sandstone itself, producing near-vertical faces rising above
straight or convexo-concave "slick rock" slopes in the same
material. Slope segmentation is usually associated with
discontinuous horizontal partings in uniform sandstone, with
the rock below the parting acting as a surrogate for a
weaker substrate. By extrapolation it is suggested that in
an arid climate the highest plane of weakness, even if only
centimeters in depth (such as that at the base of any caprock), may play a larger role in the differentiation of
cliffs and footslopes than does the more obvious contrast
between the rock types involved. Vertical faces and slick
rock slopes in massive sandstone are discontinuous, suc-.
ceeding one another both vertically and laterally. Seepage
of moisture at high-level partings appears to be a significant but not essential factor in slope segmentation, mainly
causing collapse below the seepage plane rather than above
it, to create long slick rock slopes below inconspicuous
ledges. Slick rock slopes of this type have not been
formed recently and may reflect greater moisture infiltration at some time in the past. In massive sandstone with
no exposed substrate, scarp forms alternate through space
and time as bedding discontinuities appear and die out at

different levels, so that forms are constantly readjusting
to changing physical conditions. The scarp retreat process
is spasmodic, and a time-independent steady-state development exists only in the very broadest sense. The advantage
of the principle of allometric change as opposed to the
equilibrium concept is vividly demonstrated in this landscape.

INTRODUCTION

Cuestaform landscapes in arid regions are the epitome
of weathering-limited landform systems, in which rock surfaces are kept free of waste products by the disparity between weathering rates and erosional efficiency. Cuestaform landscapes seem to exemplify the equilibrium concept
of slope development in which each rock type is associated
with a particular slope angle that equates erosional stress
to surface resistance, resulting in efficient removal of
weathering products. This appears to produce parallel rec-
tilinear slope retreat, with no evolution in form as a
consequence of the passage of time, except as entire beds
of rock are peeled away.

However, Koons (1955) drew attention to the fact that
major slope breaks commonly occur within rather than between formations, as the backwearing of cliff-forming strata in arid regions actually involves short cycles of downward cliff propagation into weaker supporting layers followed by eventual collapse due to the mechanical weakness
of the unbuttressed substrate. Each collapse only tempor-
arily returns the cliff base to the interface between the
competent and incompetent layers (Figure 1). Subsequently,
Ahnert (1960) observed that cliff forms in massive sandstones of the Colorado Plateau were highly variable, with
rounded cuesta edges in some areas, and sharp brows in
others. Ahnert suggested that climatic change had played
a role in variation of cliff morphology on the Colorado
Plateau, with pluvial spring-line sapping at the interface
between permeable sandstones and subjacent less permeable

layers producing cliff collapse and sharp cuesta edges, followed by crest rounding during interpluvial and post-pluvial
time due to reduced moisture infiltration and greater surface erosion by runoff from short-duration convective storms
(Figure 1). Ahnert commented on the limited development of
talus at cliff bases in the area he investigated, suggesting
that cliff collapse was relatively limited under present
arid conditions. In reply, Schumm and Chorley (1966) presented evidence that caprock collapse was frequent and
stressed the mechanical weakness of many poorly cemented
cliff-forming sandstones of the Colorado Plateau, which
largely disintegrate into sand during rock falls, with
larger talus blocks being quickly reduced to heaps of sand
by weathering.

All of these investigations focus upon compound or complex scarps (Schumm and Chorley, 1966) developed in alternating layers of competent and incompetent or permeable and
less permeable sedimentary rock with most morphological discontinuities being easily explained by variations in resistance, as in Koons' model, or by the possibility of contact
spring sapping, as propsed by Ahnert, with the effect of
climatic change remaining conjectural. However, intriguing
form variations also appear in simple scarps (Schumm and
Chorley, 1966) where there are no resistance or permeability
contrasts, particularly within the ubiquitous massive aeolian sandstones of the Colorado Plateau region. Form variations in apparently homogeneous rocks are puzzling when
first seen; they do not appear to have been subjected to
geomorphic analysis; and, most importantly, they are an interesting illustration of the distinctiveness of the controls of landform development in arid environments. The
present study focuses upon such morphological discontinuities within the massive cliff-forming Jurassic sandstones
that dominate the scenery in the vicinity of Moab, Utah,
but is pertinent to landscapes seen throughout the Canyonlands and Navajo section of the Colorado Plateau. These
are the same landscapes discussed by Ahnert (1960, and Schumm
and Chorley (1966).

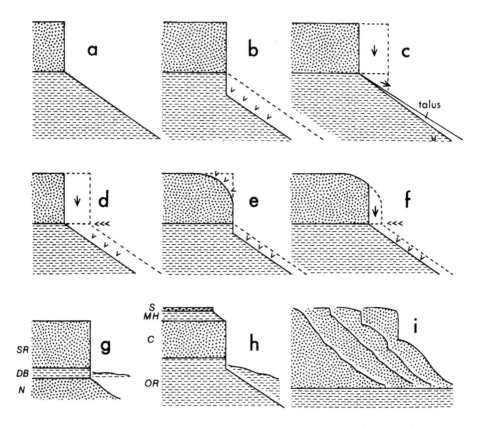

Figure 1. *Cliff retreat according to Koons (1955). A. basic cliff form with massive caprock overlying thin-bedded strata, B. downward cliff propagation by erosion on footslope, C. collapse produces talus and returns slope break to major contact. Cliff morphology according to Ahnert (1960), D. cliff retreat by pluvial contact spring sapping, E. cliff rounding by interpluvial surface runoff or depositional blanketing of spring line, F. resumption of pluvial contact spring sapping following interpluvial erosion, or removal of deposit at contact spring line. Cliff elements in study area, G. general relationships at Arches National Park: SR--Slick Rock Member of Entrada Formation; DB--Dewey Bridge Member of Entrada Formation; N--Navajo Sandstone, H. general relationships in Monument Valley: S--Shinarump Conglomerate; MH--Moenkopi and Hoskinnini shales; C--De Chelly Sandstone Member of Cutler Formation; OR--Organ Rock Member of Cutler Formation, I. representative cliff forms where only massive sandstones are exposed.*

Study Areas

This study is based on observations made during several visits to eastern Utah and northern Arizona over a period of about 10 years, principally in the regions of Arches National Park near Moab, Utah and Monument Valley, which extends into northeastern Arizona (Figure 2). However, topography similar to that which will be described exists throughout southeastern Utah, western Colorado, northeastern Arizona, and northwestern New Mexico, wherever massive sandstones are exposed. The scenery is somewhat similar throughout this area, with cuestas and detached mesas and buttes of massive sandstones of varying hues jutting sharply above weaker substrates composed of thin-bedded sandstones and sandy shales. Where such a substrate has not been exposed by downward erosion, as is frequently the case, the sandstone topography consists of joint-defined rock fins, humps, and canyon walls, displaying a variety of slope angles and slope breaks.

The vegetative cover in both the Arches and Monument Valley regions is dominated by grasses, Ephedra, Artemisia, and Juniperus, with large areas of bare ground between individual shrubs. Vegetation-free surfaces of solid rock are widespread. Evapotranspiration demands exceed the annual precipitation by an order of magnitude, with an average of less than 150 mm of precipitation occurring at Mexican Hat, about 24 km from Monument Valley, and somewhat above 200 mm at Moab. Precipitation occurs throughout the year, with a late summer-early fall maximum. All months normally have at least one day in which 2.5 mm of precipitation falls. A total of as much as 10 mm of precipitation can be expected on one day a year (in August) at Moab and does not occur during an average year at Mexican Hat. Light winter snowfalls affect both areas. For five months of the year the daily minimum temperature averages below freezing, while the normal daily maximum temperature is above 32° C (90° F) from June through August, and approaches 38° C (100° F) during July.

Lithology

In the Canyonlands and Navajo region the structure of
the sedimentary cover is undulatory and bedrock dips are usually gentle, with escarpments retreating from minor structural highs. The lack of strong and consistent dips permits
individual strata to be exposed in intricate outcrop patterns over large areas. The escarpments examined most
closely are composed of the Jurassic Entrada Sandstone in
Arches National Park, and Permian, De Chelly sandstone in
the Navajo Tribal Park. Both sandstones are massive, well-
sorted, fine-grained, rather weakly cemented by calcite and
ferric oxide, and show both aeolian crossbedding and indistinct horizontal layering (Gregory, 1917; Baker, 1936;
Schumm and Chorley, 1966; Wright and others, 1962).

The massive upper portion of the Entrada Sandstone in
the Arches region is subdivided into an upper lighter-
colored, more siliceous Moab Member and a lower buff to
salmon-colored felspathic Slick Rock Member (Wright and
others, 1962). This distinction is of no importance in the
following, as most of the cliff forms of interest are developed in the Slick Rock Member, which is 75 to 100 m thick
in the study area.

The lower 15 to 30 meters of the Entrada Formation
consist of the Dewey Bridge Member (formerly Carmel Formation; Wright and others, 1962) composed of reddish-brown
siltstone, sandy siltstone, and silty sandstone in multiple
layers 0.1 to 1.0 meters thick. The Dewey Bridge Member
overlies the massive Triassic/Jurassic Navajo Sandstone, to
which the landscape has been stripped over considerable
areas in front of the Entrada Sandstone cuestas. Cliffs in
the Slick Rock Member tend to be propagated vertically
downward into the Dewey Bridge Member, and often completely
through it to the top of the Navajo Sandstone. Even where
the Navajo Sandstone has not been exposed, there is no widespread development of Dewey Bridge footslopes--rather the
cliff foot terminates abruptly in a rolling erosional topography in the Dewey Bridge Member.

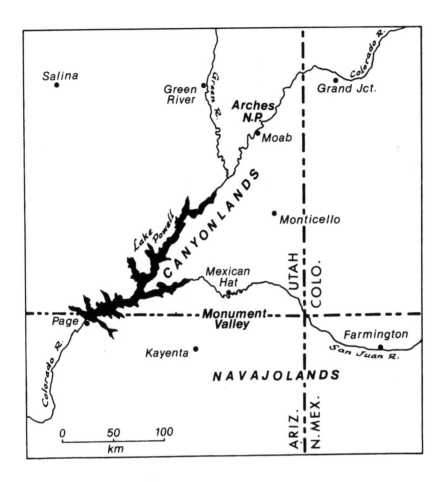

Figure 2.   Location of study area.

The Dewey Bridge Member constitutes a relatively thin
zone of weakness between the massive Navajo Sandstone and
the Slick Rock Member of the Entrada Sandstone, and is responsible for many of the natural windows and rock arches
that have caused a national park to be established here.
These have mainly been produced where basal undercutting by
erosion in the Dewey Bridge Member causes collapse of slabs
that break openings through joint-bounded rock fins in the
Slick Rock Member.   The maintenance of bold through-going

85

cliffs in the Entrada Sandstone requires exposure of, and erosional removal within, the Dewey Bridge Member. However, the converse is not true; exposure of the erodible Dewey Bridge Member does not guarantee the presence of vertical cliffs in the overlying massive sandstone.

The scenery in much of the Monument Valley region differs from that at Arches National Park mainly by the presence of imposing foot slopes below a caprock of the massive De Chelly Sandstone Member of the Cutler Formation. Thus, the region's famous 80 to 120 meter high cliffs rise above lofty pedestals composed of the darker Organ Rock Member of alternating thin-bedded (0.15 to 3 m thick) sandstones and arenaceous shales. The Organ Rock pedestals below De Chelly Sandstone cliffs commonly account for as much as half the local relief of 130 to 250 meters. As at Arches National Park, vertical cliffs have been propagated downward below the De Chelly Sandstone caprock into the less resistant Organ Rock Member, but since the latter is nearly 150 m in thickness, these extended faces are usually succeeded by the long footslopes that one rarely sees in the Arches area due to the limited depth of the Dewey Bridge Member. However, in many localities erosion has not progressed far into the Organ Rock Member and the landscape duplicates that in the Arches area.

Unlike the Dewey Bridge-Navajo contact, the transition from the Organ Rock Formation to the underlying Cedar Mesa Sandstone in Monument Valley is inconspicuous and is located some distance from the bold De Chelly Sandstone cliffs. Immediately west of Goulding, in Monument Valley, the base of the De Chelly Sandstone passes below ground level, removing the influence of the Organ Rock substrate. The subsequent discussion concentrates upon the Entrada Sandstone in the Arches region, which can serve as a model for erosional development in all of the massive aeolian sandstones in the Plateau region.

Figure 3. Cliffs in Entrada Slick Rock Member in Arches National Park; 180-degree panorama taken about 1,600 m north of Courthouse Wash. Dewey Bridge Member in foreground.

87

Figure 3 indicates the nature of the geomorphic problem
provoking the present investigation. Here it can be seen
that slopes in uniform scarp-forming sandstones are often
discontinuous, being composed of both perpendicular walls
and ramplike surfaces of varying inclinations, with ramps
and cliffs occurring both above and below one another and
alternating along the strike. Where erodible thin-bedded
substrates are not exposed, vertical sandstone walls are
nevertheless developed; conversely, where the erodible
members are exposed, the massive sandstone above may be
either cliffed or ramplike in form.

Surfaces of massive sandstone in these regions may be
resolved into several different types that are clearly differentiated, with only limited zones of transition from one
type to another. All are developed in rock of the same
physical and chemical characteristics.

Slick Rock Slope

Ramplike sandstone surfaces always consist of smooth,
multi-complex, unweathered outcrops, locally called "slick
rock" (Figure 4). The term is quite expressive of the
character of these smooth slopes, which may be found at
varying levels on sandstone scarps; occurring anywhere from
the slope foot to the slope brow. Often several discrete
slick rock slopes form vertical tiers of inclined treads on
a single staircase-like scarp composed entirely of massive
sandstone. Slick rock slopes are always composed of very
sound, clean, smooth sandstone, with little flaking evident
and no manganese stain present. The major dicontinuities
in slick rock slopes are a consequence of variations in resistance related to the indistinct horizontal bedding within
the massive sandstone. Weaknesses probably proceeding from
poorer cementation are evident in horizontal lines of
weathering pits, which I will refer to as seam lines, since
the rock appears tacked together between successive closely

spaced holes. Seam lines are always discontinuous and in the Arches region are concentrated along red bands, apparently reflecting poorer development of calcite cement within these horizons. Seam lines are often associated with minor concave slope breaks, so that slick rock walls may be terraced or multi-convex downslope as well as along the contour. Where the sandstone is broken by distinct horizontal partings, slick rock slopes normally are present below the partings, with vertical cliffs rising to varying heights from the line of the parting.

The micro-relief on slick rock slopes consists of minor indentations along vertical joint sets and shallow downslope furrows that are rather evenly spaced and that originate on the slick rock itself. The major relief features interrupting slick rock slopes are U-shaped gutters that channel drainage originating at higher levels on the scarp. Erosional removal on slick rock slopes appears to proceed in a grain-by-grain fashion by water in sheetflows that develop minimal surface roughness. Where long slick rock slopes are found below bold vertical cliffs on cuesta salients, they tend to have a narrow convex brow and a broadly concave foot. In recesses they are concave from crest to base. Thus their profiles resemble those of hillslopes in humid soil-covered landscapes.

Slick Rock Wall

Where the upper portion of a slick rock slope is interrupted by a major gutter channelling runoff from higher levels, the slick rock slopes of the gutter curve up to form vertical walls on either side. Aside from their declivity these surfaces are similar in all respects to slick rock slopes. Slick rock walls always terminate upward at a horizontal parting, above which the vertical rise is continued as a slab wall.

Figure 4. Erosional forms in Entrada Slick Rock Member, Arches National Park.

A. Slick rock slope below weathered slab wall; relief about 85 m.

B. Fresh slab wall one km north of location A; relief about 100 m.

C. *Secondary wall extended about 10 m into slick rock slope below slab wall, near location A.*

D. *Major secondary wall (center) extended about 40 m into slick rock below strongly weathered caprock.*

91

Slab Wall

In the massive De Chelly and Entrada sandstones, much of the removal is by direct collapse along joint planes or conchoidal fractures (Figures 4A, 4B). In either case undermining is a necessary prelude to the removal of a sandstone slab. The initial result of a detachment is a clean face of rock. It subsequently develops brown to black manganese stains produced by water streaming down the rock face. After staining has darkened the entire surface, or perhaps concomittantly, flaking and alveolar weathering begin to roughen the face. As weathering proceeds, flaking and the conjunction of both shallow and deep cavities remove the manganese patination, producing a much roughened exposure of rock that is more deeply stained by ferric oxide than are fresh exposures or slick rock slopes, but which are not as dark as areas of more recent detachment that retain their manganese patina.

The change from slick rock to slab wall (in any condition) is clearly defined and abrupt. Fresh slab failure removes the lower portions of slick rock slopes, but slick rock surfaces constantly tend to encroach downward at the brow of slab walls, as evidenced by the abrupt upward termination of the thin case-hardened varnished layer, which is everywhere being peeled downward from above.

Secondary Wall

Where slick rock slopes are overtopped by strongly weathered slab walls, a secondary wall has often been excavated downward into the slick rock ramp, just as cliffs grow downward into thin-bedded sandstones and shales under resistant caprocks as described by Koons (1955). This causes the concave slope break to be situated below the parting that normally separates slick rock and slab walls (Figure 4C). A parting is not indispensible to the development of a downward expanding secondary wall, for the latter can occasionally be seen developing below horizons that protrude due to superior resistance presumably related to

92

cementation or the nature of the aeolian crossbedding characteristic of the massive sandstones.

Although secondary walls in massive sandstone locally
have been excavated downward 30 or more meters, they are more
commonly but one or two meters high. The deepest excavations
are associated with small much-weathered capstones on promontories, where there is minimal load to be supported by the
secondary wall (Figure 4D). It must be emphasized that there
is no physical difference in the bedrock composing the slick
rock slope, the slab wall, and the secondary wall.

The micro-morphology of the downward growing secondary
wall is highly distinctive. Upon close examination, secondary walls show advanced chemical alteration, being covered
with fragil scales in various degrees of detachment and displaying surface efflorescences of whitish salts that are not
seen on the other surfaces. The scales at the base of secondary walls terminate in downslope-facing edges that project
over the subjacent slick rock slopes. The latter suggests:
(a) that scaling indeed occurs on slick rock slopes, with the
scales being removed by rainbeat and runoff as quickly as
they are formed; or (b) that scaling has occurred in the past
on slick rock slopes, the scales having subsequently been
removed except where protected from rainbeat and runoff by a
slab wall vertically overhead. The slab walls above secondary walls always project somewhat beyond the latter, giving
the secondary wall protection from vertically falling rain.

Colonnade Wall

Vertical faces that have weathered without collapse
for an unusually long time show a tendency to develop a columnar morphology in which weathering and erosional removal
seem to become concentrated along vertical lines, leaving
columnar septa standing out between regularly spaced recesses (Figure 5A). No horizontal change in the physical
characteristics of the rock is evident in such localities,
and the controls of this weathering pattern are not entirely clear. Vertical joints are sometimes, but not gen-

Figure 5. A. *Colonnade wall north of Courthouse Wash, producing The Courthouse; relief about 90 m.*

Figure 5. B. *Free-standing wall projecting from inset slick rock ramps with high level seepage line terminating at end of slick rock ramp; relief about 90 m.*

94

erally, visible in the recesses and possibly also occur in the septa, where some type of case-hardening may be present. A possibility is that vertical varnish streaks case-harden the rock enough to retard weathering along the line of the streaks. Removal between such streaks would eventually undercut the varnish, but leave the streaked area projecting as a septum. What can be said with certainty is that colonnade walls, where present, indicate a phase of cliff formation succeeded by an unusually long period of stability. Although colonnade walls are relatively uncommon, they are found at varying levels, some occurringhigh above the ground surface, and others reaching from mid-wall to the slope foot.

There is a clear tendency for the larger secondary walls to acquire a fluted morphology resembling the much higher colonnade walls of conjectural origin. This suggests that colonnade walls might indeed be the ultimate product of downward growth of a secondary wall. Colonnade walls display above average chemical decay in the form of flaking and alkaline efflorescences, but are less rotted than secondary walls. The difference might be attributable to the greater exposure of colonnade walls with distance below their capping entablature, which would favor more rapid removal of decayed rock.

Alveolar Wall

Occasionally, near-vertical sandstone faces are thoroughly pitted by cavernous weathering, concentrated along horizontal bands. The vertical walls were created by basal undercutting that is no longer active, as indicated by the severe alveolar weathering. Large through-going alveolar walls are confined to areas in which an actively undercut cliff base has been buried by alluvial infilling, with the stream having remained distant from the cliff since the end of the aggradational phase. Short alveolar walls also occur locally above slick rock slopes, and reflect stagnation of a slab wall above an intraformational parting.

95

Free-Standing Wall

Occasionally a fin-like, free-standing sandstone wall rises abruptly out of a slick rock slope (Figure 5B). These walls are perpendicular to the strike of the scarp, and are usually parallel to the dominant joint direction. However, neither ledges nor slick rock slopes above the point at which such walls abut the slick rock slope are broken by visible joints that could have defined the wall. Thus, walls of this type differ from the ubiquitous sandstone fins that originate where parallel joints control erosional development, as in the Fiery Furnace area of Arches National Park.

Free-standing walls meet the flanking slick rock slopes along a sharp topographic discontinuity, but are fairly thoroughly weathered. It may be that free-standing walls are less an anomaly than the long slick rock slopes flanking them for the latter are the floors of steeply-inclined troughs that terminate to either side against perpendicular cliffs normal to the strike and parallel to the free-standing wall and the dominant joint orientation. Were there evidence of large scale gravitational detachments in these areas, one could conjecture that massive slides along oblique fractures removed the rock to either side of the free-standing wall and created the slick rock slopes. In such a case, it would be difficult to explain the singular immunity to collapse of the free-standing wall itself.

Spring Lines

Evidence of spring action as a control of topographic form in the usual sense is not striking in either the Arches or the Monument Valley areas. Large and small vertical alcoves are present at the foot of slab walls, but they show no evidence of spring action. The origin of these alcoves is clear and will be discussed below. Horizontal alcoves that clearly mark spring heads are present on the up-dip sides of canyons. The most clearly evident spring line seen in the study area is at the top of high slick rock slopes bordering Courthouse Wash in Arches National Park. The rock dip on both sides is toward the wash, which follows the axis of a syncline. The active seepage planes, conspicuous for the vegetation they sustain, have little overt effect on the morphology of the slope, since they occur near the canyon rim. Minor undermining is evident and the spring line does separate long slick rock slopes from a short cliff that forms the canyon rim. However, conspicuous free-standing walls jut from these slick rock slopes, suggesting possible spring-aided slope failure (or failures) on an inclined plane, that are easily conceived as having transformed slab walls into slick rock slopes, with the free-standing walls being the last remnant of the original slope. Significantly, the slick rock slopes terminate and are succeeded laterally by slab walls wherever the spring line terminates.

The preceding discussion indicates that several inter-related morphological problems are evident in the sandstone topography of southeastern Utah.

A. The vertical sequence of slick rock slopes and slab walls in uniform massive sandstone requires explana-tion, as do lateral alternations of the sequence of forms.

B. The presence of vertical slab walls above some partings and not above others represents a problem, as does the maintenance of slab walls terminating downward in massive sandstones.

C. The mechanism of slope retreat in massive sandstone composed of alternating slick rock and slab walls is not clear.

D. The origin of colonnade walls is not clear.

E. The origin of free-standing walls *vis-a-vis* flanking slick rock slopes is not clear.

F. The possibility that spring action plays a different role than is usually proposed requires investigation.

Figure 6.   Collapse forms in Slick Rock Member above Dewey Bridge Member.

Figure 6.   A.   Collapse alcoves about 5 m high and nature of contact per-
mitting detachment; a seam line is visible at the top.

Figure 6.   B.   Recent breakdown showing initial phase of slab wall
stoping above Dewey Bridge contact; separation about 0.5 m;
note the absence of a footslope in the Dewey Bridge Member.

ORIGIN OF THE SANDSTONE CUESTAS

Before moving on to a discussion of the problems noted above, it is necessary to examine the origin of the sandstone cuestas of the principal sites studied. In both areas retreating cuestas are present because mechanically weak formations underlie massive sandstones, and have been exposed by erosion through the sandstones. This permits undermining of the resistant layer and triggers the formation and lateral retreat of perpendicular cliffs in the caprock. The degree of exposure of the weaker substrate that is required to produce such undermining is minimal. The base of the major Entrada Sandstone cuesta in the Arches region commonly exposes no more than a meter or two of the thin-bedded Dewey Bridge Member below spectacular hundred-meter-high slab walls in the Slick Rock Member (Figure 6A). The apparent rapidity of cliff retreat in the Arches region, where this contact has become exposed, is remarkable. Little downward erosion in the weak material, in front of the cliff, is apparent. North of Courthouse Wash, the cliff foot often lies in a brush-filled moat hemmed in on one side by solid rock on the other by decomposed talus and clean wind-blown sand. Furthermore, the responsibility for cliff formation does not appear to lie in the nature of the Dewey Bridge Member *per se* so much as in the presence of a single dark ferruginous shale layer some 20 to 50 cm thick, encountered one meter below the top of the Dewey Bridge Member. Removal of this friable clayey material causes collapse of the uppermost Dewey Bridge stratum, a resistant botryoidal sandstone about a meter thick. The removal of this layer in turn produces a small tensional rupture (probably along aeolian cross beds) in the overlying Slick Rock Member, followed by slight settling or a breakaway from the latter (Figure 6B). This detachment leaves a small alcove in the base of the Slick Rock Member that grows ever-larger by subsequent collapses from increasing heights. The initial tiny alcoves, often involving no more than a cubic meter of rock, activate cliff retreat by sending out slow erosional "ripples" that work across vast areas of the cliff face, producing the slab walls that reflect rapid backwearing. There is no spring action associated with these alcoves, now or in the past.

Although exposure of the Dewey Bridge/Slick Rock contact is normally followed by cliff formation in the Slick Rock Member, local exceptions exist. Some 1,600 meters north of Courthouse Wash, in the Arches area, the weaker Dewey Bridge Member is exposed in a band 800 m wide located in front of a bold cuesta, but the massive Slick Rock Member is not undercut at the base. Instead, for a horizontal distance of about 800 m the cliff foot falls midway in the Slick Rock Member, and is fronted by a convexo-concave slick rock slope as much as 60 m high that merges smoothly into erosional relief on the weaker Dewey Bridge Member (Figure 3). North and south of this locality, exposure of the Dewey Bridge Member has resulted in the formation of vertical cliffs in the Slick Rock Member in the manner described above.

The extremely prominent slick rock slope along the cuesta front in this locality resembles erosional slopes in massive sandstone where no weaker substrate has been exposed. It is instructive to follow this portion of the Slick Rock/ Dewey Bridge contact laterally from the slick rock footslope through the transition zone to the slab walls on either side. One finds that at either margin of the anomalous area, water streaming from slick rock gutters in the massive sandstone of the cuesta face has cut deeply into the Dewey Bridge Member. Plunge-pool development in the Dewey Bridge Member initiates morphological separation between the Dewey Bridge and Slick Rock members by exposing the weak shale horizon, which causes undermining and slab failure that spreads laterally from the plunge pools. The Entrada Sandstone scarp at this point has developed its peculiar morphology due to cliff stagnation following active retreat to a point at which the Dewey Bridge substrate passed below the ground level (see Figure 8E). In front of the scarp at this very point, the wash receiving runoff from the high Entrada

Sandstone cliffs tumbles into the head of a deep trench in the Navajo Sandstone below the Dewey Bridge Member. Thus, until recently, the Navajo Sandstone surface has formed a local denudational base level, preventing erosion below the Entrada Slick Rock Member. To the south, the Navajo Sandstone has been trenched by this wash for a long period of time. To the north, the structural rise out of the Courthouse Syncline brings the Dewey Bridge to so high a level that it was undercut earlier than in the anomalous section.

ORIGIN OF SLOPE SEGMENTS IN MASSIVE SANDSTONE

In the preceding location morphological separation does not occur at the contact between the thin-bedded Dewey Bridge and massive Slick Rock members, but falls midway in the Slick Rock Member itself. Figure 3 shows that where the interformational contact is not exposed, or is morphologically ineffective, vertical morphological separations in massive sandstone always occur at (or slightly below) visible intraformational partings. However, not all partings produce slope segmentation. Slope junctions occurring below partings clearly have been lowered through development of downward extending secondary walls.

Close examination of those partings that could be reached revealed the presence of three different types of bedding plane discontinuities, two of which offer enough weakness to cause minor collapse of the superincumbent rock. Partings of these latter types can be termed effective partings since they influence erosional development. As in the case of the gravitational stoping working upward from the shale stratum at the top of the Dewey Bridge Member, the initial detachment above an effective parting in massive sandstone leaves a small void that subsequently expands by feeding upon itself. The sandstone below such a parting is left as a projecting platform that is soon rounded into a convexo-concave slick rock slope (Figure 4A).

Clean partings, in which only a single plane of separation is present, fully support the superjacent rock and do

not offer the opportunity for downward displacement of any-
thing beyond small rock chips. Thus, such partings play no
significant role in cliff morphology. However, the presence
of two clean partings within a few tens of centimeters
creates a thin sheet of brittle rock that may fracture due
to the weight of the masses above it, particularly when
wetted. Closely spaced horizontal separations therefore
constitute one type of effective parting. Although it is
not evident from ground level, many intraformational discontinuities that cause slope segmentation consist of up to a
half dozen partings in a vertical space of half a meter.

The parting that creates the major slope segementation
in Figure 4A and 4C differs from both of the above. Morphological separation here is a consequence of the presence of
a mere 2 to 5 centimeters of extremely fissile ferruginous
shale that locally divides the massive sandstone into two
distinct units. In this instance an extremely minor depositional episode has sufficed to create walls rising some 50
meters above slick rock ramps of an equal height. Tiny
flakes of this very finely laminated purplish-red shale are
repeatedly encountered at the top of slick rock slopes below
high slab walls, even where the parting is too narrow and
the *in situ* shale too thin to be seen without a flashlight.
Thus, this insignificant argillaceous film creates effective
partings and is a major cliff-maker in the Entrada Sandstone.

This latter example may be instructive in the interpretation of composite scarps in general, for it indicates that
slope segmentation is an inevitable consequence of the presence of even a small amount of easily removed material below a resistant mass. Hence, the form of composite scarps
consisting of cliff-forming layers above footslopes in thin-
bedded or less resistant material are less a reflection of
the physical dissimilarity between the materials above and
below the slope break than they are an indication that a
zone of weakness occurs below a uniformly resistant mass.
Sandstone separated from sandstone by an effective parting
will create a composite scarp as surely as sandstone over

a thick mass of shale. It is not the mass of the shale that creates the footslope, but the first fissile bed below the capping mass. What lies below this matters little as long as it is not more resistant than the higher cap-rock.

NATURE OF SCARP RETREAT IN MASSIVE SANDSTONE

In Figure 3 it is clear that scarp morphology varies drastically along the front of the Entrada Sandstone cuesta. This is a consequence of the domination of scarp morphology by the localized nature of effective partings in the massive sandstone. In the cliff-forming sandstones of the Colorado Plateau, effective intraformational partings are rarely continuous over any great distance. Rather they appear to exert their morphological influence over a limited area (usually less than a square kilometer) then disappear as other partings take over their role, but at different levels (Figure 7). Thus we may see the beginnings of slab wall development at the expense of slick rock slopes as effective partings appear, and, in the same view, cliff stagnation as other partings close, allowing the crests of the slab walls they have created to be rounded and incorporated into higher slick rock slopes or to degenerate into alveolar walls. Although the base of the massive sandstones are not exposed, preventing undermining and the creation of through-going slab walls, scarp retreat continues due to collapse above effective partings. Hence, the retreat progresses in a discontinuous piecemeal fashion, with slab walls constantly being spawned, growing, and dying at varying levels, as diagrammed in Figure 8.

Seam lines, reflecting poorer cementation, play a very minor role in scarp retreat, interrupting slick rock slopes and occasionally producing continuous indentations that over-steepen slopes without creating true slab walls. Nowhere have collapses been observed to take place above seam line indentations; however, where effective partings close, they commonly grade into seam lines.

103

Figure 7. Discontinuous effective partings within Slick Rock Member; relief about 85 m.

Figure 7 A. Morphological changes due to terminations of effective partings.

Figure 7 B. Compound slick rock slope reaching to effective partings at different levels on scarp face.

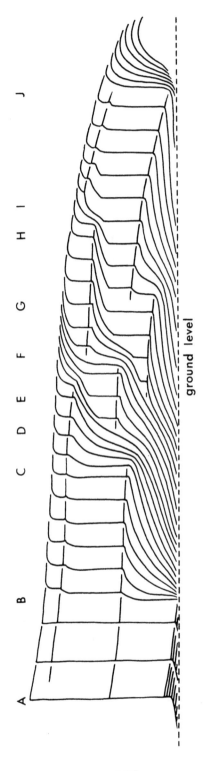

Figure 8. Development of scarp forms in massive sandstone through time and space. For simplicity a constant ground level (dashed line) is assumed during scarp retreat, along with equal thicknesses of removal from major slab walls in each unit of time. At A a through-going cliff is present due to sapping above a thin-bedded substrate. At B the thin-bedded substrate passes below ground level, scarp retreat slows, and effective intraformational partings assume control of scarp form. Effective partings close at C, E, F, G, H, and J, leading to local slab wall stagnation and rounding into slick rock. Partings that open at D, E, F, G, and I initiate growth of new slab walls. Note that the effect of former partings in rocks that have been removed continues to be expressed in the form of slick rock ramps and concave slope breaks. Lowering of ground level during backwearing would cause cliff extension upward from major contact.

Decrease in the size of capping masses reduces parting effectiveness. Relatively small capping masses tend to be rounded or honey-combed by surface weathering processes, and show minimal evidence of recent detachments (Figure 4D). It could be expected that the rate of rock failure would diminish where a smaller mass is supported above a zone of weakness. The secondary walls that grow downward below the caprock following each retreat of the caprock face substantiate this. The highest secondary walls always are developed below very small caprocks. The downward removal in these instances creates considerable pockets in the slick rock slopes below. The last detachments to have occurred from some of the smaller capping masses have initiated a new generation of secondary walls working downward into older secondary walls, a phenomenon never seen below a massive caprock.

Finally, recent initation and upward growth of slick rock slopes from ground level rather than below a parting may be seen where alluvial infilling has buried cliff bases. This is a fairly common phenomenon due to the general phase of alluviation that prior to the 1880's created level infills throughout the Colorado Plateau region. Courthouse Wash in the Arches area is presently bordered by a series of paired fill terraces rising at least 10 meters above the current broad alluvial channel. The two highest fills have caused the stagnation of cliffs cut into the Entrada Slick Rock Member by the lateral planation of the stream. The most striking of these stagnant cliffs is the much-weathered and possibly secondary (downward-extended) colonnade wall north of Courthouse Wash that doubtless provided the latter its name (Figure 5A). Undercutting by arid weathering and erosion processes at the new ground surface has girdled the foot of this wall, creating an indention 0.3 to 1 m deep and 2 to 3 m high. Included are some niches excavated as much as 2 m into the rock face. The floors of this and similar indentations slope outward, and are narrow but fairly typical slick rock slopes, exposing sound rock. The backwalls are veneered with partially detached flakes. It

106

appears that deepening of the basal niches has already caused some failures of the overlying massive but weakly cemented sandstone and has removed column bases and initiated areas of slab wall that have themselves been indented by basal weathering. This indicates a rapid rate of cliff retreat, and shows that slick rock slopes and slab walls can develop in the absence of any form of horizontal parting. However, the lack of permanence of the cliff foot under such conditions suggests that forms initiated in this manner could not be expected to have any major or long lasting effect on scarp morphology, for extension of the new slick rock slope across effective intraformational partings would transfer subsequent control of erosional form to the partings, producing slick rock and slab wall sequences similar to those encountered wherever a weaker substrate has not been exposed.

RELATION OF SLICK ROCK SLOPES TO DETACHMENTS

The convexity at the crest of many slick rock slopes probably has little to do with the amount or mode of water flow over these slopes (Figure 4A, 4C). These convexities are the normal result of detachments from higher slab walls, which simply leave a shelf at their base (below the effective parting). The brink of this shelf undergoes rounding and lowering by normal weathering and removal processes, but is not driven back as rapidly as the slab wall above it. Thus, especially at the tips of scarp promontories where small removals can cause large cliff retreats, the convex crests of slick rock slopes thrust boldly outward from the foot of higher slab walls. Where water streams onto slick rock slopes from slab walls above them, as along the flanks of salients, the erosional stress on the slick rock slope is greater and the upper convexity is only weakly developed below a secondary wall, or may not be present at all.

In several locations, long straight slick rock slopes extend upward almost all the way to the cuesta brow. These are the steep ramps previously noted as a form associated

with free-standing walls and spring lines (Figure 5B).
Whereas convexo-concave slick rock slopes tend to be smooth
except for seam lines and scattered tafoni development,
many of the straight through-going ramps have been considerably roughened by tafoni development, and appear different
in origin from the slick rock slopes flanking scarp promontories and those recesses that show no evidence of spring
action.

Rather than undergoing slow but continuous erosional
development as slab walls retreat above them, these steeply
pitching flat-floored ramps have the appearance of having
been made in a short time by one process, then modified over
a long time by other processes. Inasmuch as such slopes
always appear in association with active seepage lines, or
lie below lateral extensions of the planes from which seepage is occurring, an origin by way of large scale collapses
along inclined fractures seems likely. These fractures
would be similar to those deduced by Koons (1955) as the causes
of collapse of walls propagated down into incompetent material below a massive caprock. Although Koons indicated that
such fractures might be difficult to observe, since their
development should be followed very shortly by detachment
along them, at least one example can be seen in the Arches
region, cutting the Dewey Bridge Member along the entrance
road to the park. It is visible because a collapse has already occurred, removing some, but not all, of the material
resting on it.

In the case of the massive Entrada Sandstone, the high
porosity but low permeability of this formation, pointed out
by Schumm and Chorley (1966), would seem to favor the development of seepage planes along effective partings near the
surface of the mass but not deep within it, with wetting
of the rock at shallow depth under seepage planes, but not
far below them. A decrease in internal friction by wetting
below a seepage plane could easily be imagined as producing
failure there, along an inclined plane, rather than the
spring line undercutting usually stressed by investigators
of cuesta scenery (e.g., Ahnert, 1960). Collapses along

inclined planes in such settings would create long slopes
and short high level cliffs rather than the major through-
going cliffs that have been regarded as products of spring-
line sapping at the base of the massive layer. The resulting
short cliff would be confined to the surface stratum that
has open joints which feed water from the plateau surface to
the seepage plane, with the steeply inclined failure surface
developing below the seepage plane. This is the configuration associated with the abrupt free-standing walls and inset slick rock ramps mentioned earlier. The maximum projection plane of free-standing walls in the Arches region seems
to be an indication of the position and form of the scarp
face prior to the series of collapses that isolated them.
Their flanks appear to be little modified planes of separation along latent joints that may not be visible within the
mass of the Entrada Slick Rock Member, but whose trend is
clearly expressed on the plateau surface (Figure 9).

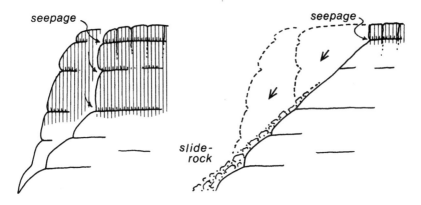

Figure 9. Hypothesis in explanation of the relationship between seepage
planes, free-standing walls, and inset slick rock slopes.
Vertical lines on left diagram show depth of wetting of
massive sandstone during pluvial phase, with several seepage
planes possible. Wetting decreases strength of sandstone,
causing failure along inclined planes within the wetted zone.
Present landscape (right) shows relict slide planes with
local slide rock and existing high level seepage plane.
Free-standing walls are left between detached masses.

The absence of evidence of deposits resulting from the collapses deduced above is significant. Since such collapses should have involved large masses moving on an inclined plane rather than free falls from high slab walls, the resulting debris should not have disintegrated on impact. Yet neither slide debris of an appropriate type nor recently-formed, inclined, detachment surfaces are conspicuous in the Arches region. The erosional cliff bases in the area of maximum seepage and deduced rock sliding along Courthouse Wash are not visible due to alluvial infilling, which has drowned an unknown portion of the erosional relief. However, the weathered appearance of many inset slick rock ramps suggests that if they are detachment planes they are relatively ancient features, with associated deposits having long since disintegrated to sand.

CLIMATIC CHANGE AND RELICT LANDFORMS

If the flat-floored slick rock ramps are indeed relict forms, they would seen to reflect a time of greater moisture availability, producing increased activity along high-level seepage planes. If these inset ramps have been correctly interpreted here, they are the only form in the region that cannot be explained in terms of the present climate and processes that seem to be active today.

The forms observed by Ahnert, consisting of cliffs eating into rounded (slick rock) slopes, and rounding (slick rock development) of cliff brows, do not seem to require a climatic change. Brink rounding and cliff-foot undermining go on constantly. Collapses are momentary interruptions that break into undermined slopes that have been undergoing rounding since the previous collapse. Thus, both slope types described by Ahnert seem to evolve simultaneously. Basal sapping could increase or decrease due to periodicity in spring-line sapping, as Ahnert proposed, or could merely indicate lateral changes in the effectiveness of intraformational partings and the specific nature of the contacts between formations, which are the most important

110

factors in the development and maintenance of bold cliffs.

However, the picture is not an altogether straightforward one. In the Monument Valley region one sees excellent examples of cliff stagnation due to sand deposits that blanket the contact between the De Chelly and Organ Rock formations, seeming to substantiate Ahnert's hypothesis, but mere blanketing of the contact is enough to produce the same result, regardless of the presence or absence of spring action. Nevertheless, there seems to be evidence that spring sapping in the Arches region may be reduced in scale today as against some time in the past, as Ahnert would have imagined, although the nature and effect of spring action seem quite different than presumed by Ahnert. While Schumm and Chorley's findings (1966) vitiate the climatic change hypothesis and while the present analysis indicates that strongly contrasting scarp forms can develop without climatic change, the possibility that climatic change can be seen in these landscapes cannot yet be completely excluded. Every vertical or lateral variation in cliff form can be explained as the result of a temporal change in climate or a spatial change in parting effectiveness. The present study suggests that the latter is adequate to account for most inter- and intraformational form variations encountered in the Plateau Country, but perhaps not all of them.

CONCLUSION

In the above, the focus has been upon the fact that in an arid weathering-limited landscape a very minor intraformational discontinuity can cause uniform massive sandstone to develop differing morphologies similar to the morphological contrasts between massive and thin-bedded sedimentary rocks. In the massive sandstones the lower unit, although physically indistinguishable from the higher, mimics the behavior of a thin-bedded substrate, producing footslopes that are encroached upon by secondary walls extended downward below the cliffed upper unit, which plays the role of a caprock. The principal morphological difference between

sandstone over sandstone along discontinuous intraformational partings and resistant caprocks over less competent substrates are lateral changes in scarp form in the massive
sandstone due to the opening and closing of the morphologi-
cally-effective partings. In massive sandstone with no
exposed substrate, perpendicular slab walls and ramplike
slick rock slopes of every inclination succeed one another
through space and time as intraformational partings appear
and die out at different levels, initiating and damping out
the potential for sapping and upward stoping from minor
collapse features. Thus, slope forms in massive sandstone
bear no consistent relationship to the physical properties
of the mass of the rock of which they are composed, and are
in a constant state of readjustment as opportunities for
sapping and collapse wax and wane.

Under these conditions cliff retreat must be fitful
from place to place and time to time, with exposure of
effective partings causing rapid local backwearing, and clo-
sure of these partings causing rapid local cliff stagnation.
The appearance of stagnation, consisting of through-going
slick rock slopes, may, in fact, proceed from a period of
maximum activity in which cliff collapses were occurring
along inclined fractures developed in wetted rock below
seepage planes.

Continuous form adjustments due to the structural pec-
uliarities of massive aeolian sandstones and their sensitivity to changing moisture inputs make the study of massive
sandstone scarps an exercise in the application of the
principle of allometric change (*cf*, Bull, 1976). As each
of the many massive sandstones of the Colorado Plateau were
exposed they initially produced, over their entire area of
outcrop, landforms of the general type described here. Only
with the exposure of the continuous plane of weakness at
the base of each massive sandstone did the importance of
inheritance and continuous piecemeal evolutionary change
become irrelevant to the local landform system. The opportunity for sapping at the top of a thin-bedded substrate far

112

exceeds that associated with discontinuous partings within massive rock, and results in regular cliff retreat according to the Koons model. But due to the undulatory geological structure of the Plateau Country and the proportion of massive sandstones in the sedimentary column, at least two general models are necessary to account for the regional landforms; one with weaker substrates exposed and one in which they are not. Beyond this, the distinctive structural characteristics of the many stratigraphic units involved produce a host of local variations on each of these general models.

REFERENCES CITED

Ahnert, F., 1960, The influence of Pleistocene climates upon the morphology of cuesta scarps on the Colorado Plateau: Ann. Am. Assoc. Geog., v. 50, p. 139-156.

Baker, A. A., 1936, Geology of the Monument Valley-Navajo Mountain region, San Juan County, Utah: U. S. Geol. Survey Bull. 865, 106 p.

Bull, W. B., 1976, Allometric change of landforms: Geol. Soc. America Bull., v. 86, p. 1489-1498.

Gregory, H. E., 1917, Geology of the Navajo country: U. S. Geol. Survey, Prof. Paper 93, 161 p.

Koons, D., 1955, Cliff retreat in the southwestern United States: Am. Jour. Science, v. 253, p. 44-52.

Schumm, S. A. and Chorley, R. J., 1966, Talus weathering and scarp recession in the Colorado Plateaus: Zeitschr. fur Geomorph., N. F., v. 10, p. 11-35.

Wright, J. C., Shawe, D. R., and Lohman, S. W., 1962, Definition of members of Jurassic Entrada Sandstone in east-central Utah and west-central Colorado: Am. Assoc. Petrol. Geol. Bull., v. 46, p. 2057-2070.

TECTONIC GEOMORPHOLOGY NORTH AND SOUTH
OF THE GARLOCK FAULT, CALIFORNIA

William B. Bull
Leslie D. McFadden

Geosciences Department
University of Arizona

ABSTRACT

Five differential equations that interrelate uplift, erosion, and deposition along stream systems that cross the mountain fronts of the northern Mojave Desert were used to appraise three classes of Quaternary tectonism. Class 1 (active tectonism) terrains are characterized by mountain-front sinuosities of 1.2 - 1.6, unentrenched alluvial fans, elongate drainage basins with narrow valley floors and steep hillslopes even in soft materials. Class 2 (moderate to slightly active tectonism) terrains are generally character-ized by mountain-front sinuosities of 1.8 - 3.4, entrenched alluvial fans, large drainage basins that are more circular than class 1 basins in similar rock types, steep hillslopes and valley floors that are wider than their floodplains. Class 3 (tectonically inactive) terrains are characterized by mountain-front sinuosities of 2 to 7, pedimented mountain fronts and embayments, steep hillslopes only on resistant rock types, and few large integrated stream systems in the mountains.

Marked contrasts of landscapes, that are due to differ-ent relative rates of base level fall, are present north and south of the strike-slip Garlock fault. In a northern sub-area, class 1 terrains generally occur on the west sides and class 2 terrains on the east sides of the eastward tilted

The views and conclusions contained in this document are those of the authors and should not be interpreted as necessarily representing the official policies, either expressed or implied, of the U. S. government. Sponsored by the U. S. Geological Survey No. 14-08-0001-G-394.

fault - block mountains. Extreme tectonic stability is shown
by the dominance of class 3 terrains south of the fault.
Pedimentation has left only remnants of the formerly more
extensive mountain ranges and their drainage basins. Between
the extremes of these two subareas is a transitional subarea north of the Garlock fault, where the magnitudes of
Quaternary uplift of the mountain fronts decrease towards
the south.

INTRODUCTION

The hillslope, stream, and depositional subsystems of
arid fluvial systems tend toward configurations that are
the result of the interaction of many variables. One important independent variable is base level change. Tectonic
base level fall partly determines the relief of a fluvial
system, and a pulse of uplift along a mountain front will
cause adjustments in the systems that flow across the front.
For example, uplift can change a stream's erosional work
from predominately lateral cutting to downcutting, which will
leave the former floodplain as paired strath terraces. Of
course, one also has to take into account the concurrent
effects of other independent variables, such as climate
(which varies with time) and rock type and structure (which
varies with space). These independent variables largely determine the morphologies of hillslopes, valleys and drainage
nets; and affect processes pertaining to soil and vegetation,
sediment yield, and discharge of water and sediment from the
erosional part of the system.

The subject of tectonic geomorphology deals with (1)
the impact of tectonic base level fall on the processes
and morphologies of fluvial systems and (2) assessment of
the relative degrees of tectonic activity of mountain fronts,
or other structural elements, during the Quaternary. These
are the chief subjects of a book manuscript by W. B. Bull
entitled, Climatic and Tectonic Geomorphology of Arid
Fluvial Systems.

This article deals with the second type of study. We wish to report the results of an application of a tectonic geomorphology model to the landscape north and south of the Garlock fault in eastern California. This part of the Basin and Range Province (Figure 1) has some obvious differences in relief and continuity of mountain blocks. The purpose of this article is to quantify some of these differences in mountain fronts, and to evaluate areal variations in Quaternary tectonism. The scope of the article includes a brief outline of the tectonic geomorphology model, discussion of two landscape parameters that are useful in defining relative rates of uplift, and classification and discussion of the relative rates of Quaternary mountain front uplift in the study area. Uplift and/or erosion of the mountains during the Quaternary (the last 1.8 m.y.) accounts for most of the configuration of the landscape elements discussed in this article. Only major, localized tectonic processes will be considered. Broad warping and minor faulting of only a few meters are not easily accommodate by the model and only those mountain fronts that are more than 10 km long are analyzed. A mountain front is considered as being a zone that includes part of the mountains, the adjacent escarpment and part of the basin adjacent to the escarpment. The fronts studied are show in Figure 1 and the general rock types are shown in Figure 2.

The names of the mountains on Table 2 are not identical with geographic place names because the mountain fronts were selected for this study on the basis of topographic, lithologic, geomorphic and structural, continuity. For example, the Spangler Hills occur just north of the Garlock fault, but are part of the Argus Range structural-lithologic block. This block is named after the largest mountain mass within it and the fronts are numbered from north to south on both the east and west sides of the block.

The geomorphic characteristics of the mountain fronts do not change abruptly at the Garlock fault. Because of the transitional changes that are characteristic of the differ-

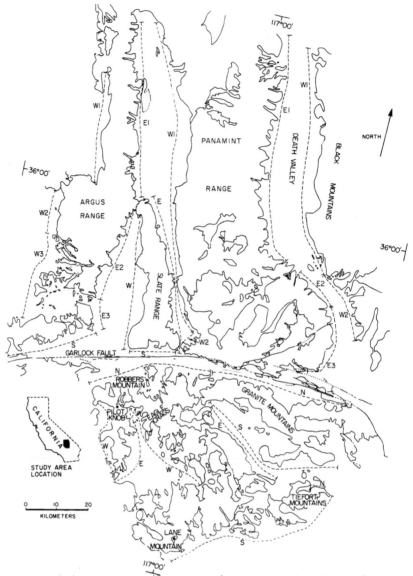

Figure 1. Locations of the larger mountain ranges in the study area, and the mountain fronts analyzed.

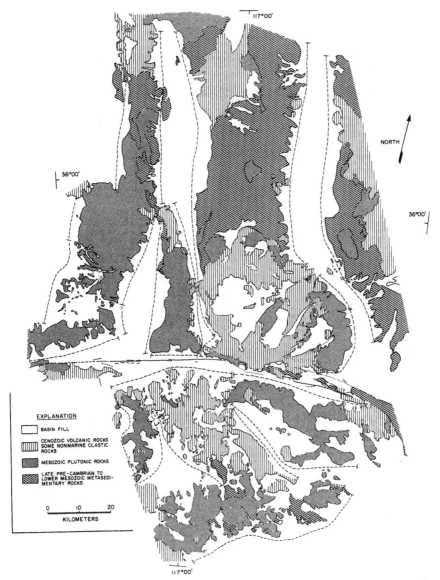

Figure 2. *Generalized geology of the study area, from Jennings and others (1962) and Jennings (1958).*

ent blocks, northern, transitional and southern subareas were separated for study and comparison (see Table 2).

The basins and low ranges receive less than 100 mm of precipitation during the average year, but the 3200 m high Panamint Range locally receives more than 300 mm. Most of the precipitation comes during the winter, although infrequent, but the intense rains from convective storms during the hot summers account for some of the major streamflow events.

Earlier Studies

The eastward striking Garlock fault in the Mojave Desert has a left lateral displacement of at least 48-64 km (Smith, 1962; Smith and Ketner, 1970; Davis and Burchfiel, 1973) and separates markedly different landscapes to the north and south. "South of the Garlock is the Mojave Desert block of generally old, low mountain blocks, largely buried by basin deposits in the west. North of the Garlock fault is a region of high and exceedingly active fault-block ranges, the Sierra Nevada, Argus, Panamint, and Black Mountains" (Hamilton and Myers, 1966). Davis and Burchfiel conclude that the Garlock fault is an intracontinental transform structure that is related to the extensional origin of the Basin and Range Province north of the fault. Both strike-slip and vertical tectonic movements have occurred north of the Garlock fault. Wright and Troxel (1967) describe less than 8 km of right-lateral movement along the Death Valley fault zone, and Smith (1975) found that the right-lateral displacement of Quaternary landforms by the Panamint Valley fault zone exceeds the vertical offset of the landforms.

THE EFFECTS OF VERTICAL TECTONIC MOVEMENTS
ON FLUVIAL SYSTEMS

A process that changes the altitude of a point on a streambed is a local base level process. Base level processes include stream-channel downcutting in the mountains

120

(w) and erosion (e) or deposition (s) on the piedmont adjacent to the escarpment. These three interdependent processes are affected by the relative uplift (u) of the mountain front. The four base level processes noted above affect the processes and morphologies of the stream and hillslope subsystems, the loci of erosion and deposition and, therefore, the topography of the basins.

The effects of continued rapid uplift of mountains relative to the adjacent basins, by either continuous or pulsatory uplift, result in a distinctive suite of landforms. Figure 3 depicts the accumulation of thick alluvial fan deposits adjacent to a faulted mountain front. Either channel downcutting in the mountains and/or basin deposition will tend to cause the stream channel to become entrenched into the fan apex, which will shift the loci of fan deposition down the fan. Counteracting the tendency to trench the fanhead is the uplift of the erosional subsystem along the range-bounding fault. Continued channel downcutting will occur only when uplift equals or exceeds the sum of the two local base level processes that are tending to cause the fanhead trenching as shown by equation 1:

$$\Delta u/\Delta t \gtreqless \Delta w/\Delta t + \Delta s/\Delta t \qquad (1)$$

Equation 1 is but one of five equations interrelating local base level processes for three different tectonic environments. The equations are shown in Table 1, and they form the basis of three classes of relative tectonic activity of mountain fronts within a given study area during the Quaternary.

The assignment of a tectonic activity class for a given mountain front is based on numerical parameters that describe diagnostic landform morphologies within both the erosional and depositional subsystems. The typical landform of Table 1 is but one of many diagnostic landforms for a given class. For example, class 1 landscapes, when compared to class 3 landscapes of similar total relief, climate, and rock type, have more convex ridgecrests, steeper footslopes, narrower and steeper valleys, less sinuous

121

mountain fronts, thicker basin deposits next to the mountains, and minimal soil profile development on the piedmont (Bull, 1973). Thus, the tectonic geomorphology model defines the interrelations of tectonic base level change to the other local base level processes by the equations of Table 1, and allows the assignment of an appropriate tectonic activity class by quantitative descriptions of selected landscape elements.

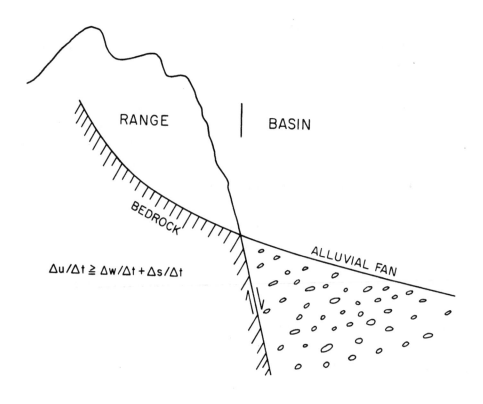

Figure 3. Interrelations of local base level processes conducive for the accumulation of thick alluvial fan deposits next to a mountain front.

SELECTED PARAMETERS THAT DESCRIBE THE INTERACTION OF
FLUVIAL SYSTEMS AND VERTICAL TECTONIC MOVEMENTS:
THEORY AND RESULTS

For the purposes of this reconnaissance, two chief parameters were used that are reliable and easy to obtain.

Mountain-front sinuosity uses the two horizontal dimensions of the landscape at the junction between the erosional and depositional subsystems. The valley floor width-valley height ratio uses one vertical and one horizontal dimension at a given point along the stream in the erosional subsystem, and, thus, is affected by variables and perturbations both upstream and downstream from the measurement point.

Mountain-Front Sinuosity

Many workers have used qualitative aspect of mountain fronts to make general assessments of the degree of tectonic activity present. A straight mountain front might be indicative of an active fault or fold while an embayed, pedimented front generally is considered to be representative of tectonic quiescence. Because the plan views of most faults and folds are gently curving or straight, the degree of erosional modification of tectonic structures can be measured by a mountain-front sinuosity index. Mountain-front sinuosity (S) is the ratio of the length along the edge of the mountain-piedmont junction (Lmf) to the overall length of the mountain front (Ls) as shown in equation 2.

$$S = \frac{Lmf}{Ls} \tag{2}$$

The sinuosity of a mountain front at any given time is a balance between the tendency of uplift to maintain a fairly straight front that coincides with a tectonic structure and the work of streams that tend to erode irregularities into the front. Erosional retreat of the front from the range bounding tectonic structure also increases with time and is accompanied by increasing values of S.

Values of S are in part a function of the scale and detail of the maps and/or aerial photos. Small-scale topographic maps (1:250,000) provide only crude estimates of S, but aerial photographs, large scale (1:62,500) topographic maps, and 1:250,000 geologic maps generally are adequate. Results obtained from the different sources of data generally are about the same, but in some cases aerial photos or small scale topographic maps will give values of S that are 20-30%

123

higher than the 1:250,000 geologic maps. Geologic maps that depict intensive pedimentation cannot be used.

Studies made elsewhere in the Sonoran and Mojave deserts show that each of the tectonic activity classes of Table 1 has fairly distinctive ranges of mountain-front sinuosity. Class 1 fronts generally range from 1.0 to 1.6, class 2 from 1.4 to 3, and class 3 from 1.8 to more than 5. Sinuosities of more than 3 characteristically are associated with mountain fronts that are so embayed and pedimented that the range-bounding tectonic structure initially responsible for the mountain front may be more than 1 km from the present erosional front.

Class of Relative Tectonic Activity	Local base level processes: u, uplift; w, channel downcutting; s, piedmont deposition; e, piedmont erosion; t, time.	Typical landform
1. Active	$\Delta u/\Delta t \gtreqless \Delta w/\Delta t + \Delta s/\Delta t$	Alluvial fan that is receiving deposits on fanhead
2. Slight	$\Delta u/\Delta t < \Delta w/\Delta t > \Delta e/\Delta t$	Entrenched alluvial fan with old soils on fanhead
3a. Inactive	$\Delta u/\Delta t << \Delta w/\Delta t > \Delta e/\Delta t$	Dissected pediment or pediment terraces
3b. Inactive	$\Delta u/\Delta t << \Delta w/\Delta t = \Delta e/\Delta t$	Undissected pediment
3c. Inactive	$\Delta u/\Delta t << \Delta w/\Delta t < \Delta e/\Delta t$	Undissected pediment

Table 1. *Mountain Front Tectonic Activity Classes (from Bull, 1973).*

The sinuosities for the mountain fronts of the study area are summarized in Table 2. They range from 1.2 to 7.2. The mean mountain-front sinuosities are 2.1, 4.1 and 3.8 for the northern, transitional and southern subareas respectively and the difference between the northern and southern sub-

Mountain Front and Direction Front Faces	Mountain-Front Sinuosity	Mean Valley Morphologies Valley Floor Width/Valley Height Ratio (V_f)	Tectonic Activity Class*
NORTHERN SUBAREA			
Black Mountains W1[1]	1.4	0.2[3]	1
Black Mountains W2[1]	3.1	2.0	2
Panamint Range E1	3.4	1.1	2
Panamint Range E2	2.9	1.5	2
Panamint Range W1[1]	1.6	0.055[3]	1
Slate Range E	1.8	2.3	2
Slate Range W	1.4	0.3[3]	1
Argus Range E1[1]	2.5	3.6	2
Argus Range E2[1]	2.0	2.0	2
Argus Range W1[1]	1.2	0.9	1
Argus Range W2[1]	1.5	0.5[3]	2
TRANSITIONAL SUBAREA			
Panamint Range E3	3.5	8.4	3
Panamint Range W2	2.9	[2]	3
Slate Range S	4.0	3.0	3
Argus Range E3[1]	4.1	6.2	3
Argus Range W3[1]	7.2	2.5	3
Argus Range S[1]	2.6	10.0	3
SOUTHERN SUBAREA			
Granite Mountains N	2.7	3.1	3
Granite Mountains S	3.4	4.0	3
Tiefort Mountains S	3.0	3.7	3
Eagle Crags E	1.8	2.1	3
Eagle Crags W	6.0	47.0	3
Robbers Mountains N	6.5	2.9	3
Pilot Mountain E	3.4	3.9	3
Pilot Mountain W	3.6	12.0	3

[1] *Mountain-front sinuosities derived from 1:250,000 scale 1° x 2° California Division of Mines Geologic Sheet.*

[2] *No major drainage basins along mountain front, therefore no V_f values obtained.*

[3] *V_f value is a maximum, width of valley floor <100 meters.*

* *See Table 1.*

Table 2. Geomorphic Data for the Mountain Fronts

areas is signficant at the .005 level (i.e. α = .005). The mountain front is considered to coincide with the range-bounding geologic structure only for the class 1 fronts. For the class 3 fronts of the southern subarea, there are few geomorphic clues as to the locations of possible range-bounding structure that have been tectonically active in the past.

Valley Floor Width - Valley Height Ratio

Obvious differences in the transverse morphologies of valleys, such as v-shaped canyons and broad-floored pediment embayments can be described by a simple ratio, the valley floor width-valley height ratio. At a given distance upstream from the mountain front (1 km in this study), comparison of the width of the floor of the valley with the mean height of the valley, provides an index that indicates whether the stream is actively downcutting (being dominated by the influence of a base level fall at some point downstream) or is primarily eroding laterally into the adjacent hillslopes. If we let V_{fw} be the width of the valley floor, and E_{ld}, E_{rd}, and E_{sc} be the altitudes of the left and right divides and the stream respectively, then the valley floor width-valley height ratio (V_f) is defined as,

$$V_f = \frac{V_{fw}}{\frac{(E_{ld} - E_{sc}) + (E_{rd} - E_{sc})}{2}} . \qquad (3)$$

The location of the cross-valley transect within a drainage basin affects the values of V_f. Valley floors tend to become progressively narrower upstream from the mountain front and for a given stream system the values of V_f tend to become progressively larger downstream from the headwaters. In this study the transects for determining V_f were located 1 km upstream from the mountain front in the larger drainage basins for a given mountain range. The reason for working with a selected size range of drainage basins is that smaller streams tend to maintain the downcutting mode of operation longer than the more competent larger streams.

126

The dimensions for V_f vary in their ease of measurement.
Detailed topographic quadrangles provided accurate determin-
ations of the altitudes of the divide above the stream
channel, but the width of the valley floor could not be
measured with equal accuracy, even where well defined by the
topographic maps. In the narrow canyons that are typical of
class 1 mountains, only maximum values of valley floor width
could be estimated, which resulted in maximum values of V_f.
However, even when one uses conservative methods, such as
maximum values, highly significant differences in V_f are
present for the suites of mountain ranges north and south
of the Garlock fault.

The valley floor width-valley height ratios are summarized in Table 2. They range from about 0.05 to 47. The
mean ratios are 1.3, 6.1, and 11.0 for the northern transitional, and southern subareas, respectively; and the difference between the northern and southern subareas is significant at the .01 level. In the canyons of the class 1 landscapes the floodplain width and the valley-floor width are
the same, but in class 2 valleys the floodplain width is less
than the valley floor width. In class 3 terrains the ratio
describes pediment embayments of the broad valleys upstream
from the embayments. Drainage basins were best defined for
class 1 fronts, while along some class 3 fronts, the
scattered inselbergs so poorly defined mountainous drainage
basins that no ratios were computed (see Figures 4 and
6). For most of the fronts, ratios were computed for one or
two drainage basins, and for Robbers Mountain (north), three
measurements were made.

Drainage Basin Shape

The lack of well defined drainage basins in some of the
class 3 terrains precluded extensive comparisons of drainage
basin shape, but basin shape was compared for those drainage
basins that extend from the divide to the mountain fronts
on the east and west sides of the central Panamint Range.

127

Figure 4. Class 1 mountain front on the west side of the Panamint Range at Redlands Canyon.

The migration of drainage basin divides that occurs as
adjacent basins compete for space results in planimetric
shapes that can be described quantitatively. The typical
basin of a tectonically active mountain range is elongate
(Davis, 1909, p. 343) and basin shapes become progressively
more circular with time after cessation of mountain uplift.
The planimetric shape of a basin may be described by an
elongation ratio (Cannon, 1976).   (4)

$$Re = \frac{\text{diameter of circle with the same area as the basin}}{\text{distance between the two most distant points in basin}}$$

Although the Panamint Range is a block that has been tilted
eastward (Maxson, 1950; Hunt and Mabey, 1966; Drewes, 1963;
Denny, 1965 and Hooke, 1972), the drainage divide between
the east and west flowing streams is in the central part of
the range. The western margin of the range is a class 1
front, and the eastern margin is a class 2 front. The elongation ratios range from 0.39 to 0.65 (mean is 0.53) for the
west-side basins, and from 0.66 to 0.74 (mean is 0.69) for
the east-side basins. The difference between the means is
significant at the .01 level. Drainage basin widths are much
narrower near the mountain front on the tectonically active
west side where the streams have directed much of their en-
ergies to downcutting. Apparently the lack of continuing
rapid uplift along the east side has permitted widening of
the basins upstream from the mountain front and the production of well-developed straths that extend far into the
mountain range along the main streamcourses.

QUATERNARY TECTONIC ACTIVITY OF THE MOUNTAIN FRONTS

The mountain fronts of the study area were assigned to
appropriate relative tectonic activity classes (see Table 1)
on the basis of, (1) the mountain-front sinuosities and
valley floor width-valley height ratios, and (2) other features (such as entrenched or unentrenched alluvial fans)
which were observed during trips to the study area and on
flights over the area. All of the mountain fronts are considered to be class 3 except in the northern subarea where

four class 1 and seven class 2 fronts are present.

A typical class 1 landscape is shown in Figure 4. The rugged, narrow drainage basins have V-shaped cross-valley profiles and the valley floors are the same width as the floodplains. The canyons are being downcut into the Paleozoic sediments and are notched into an eroded fault scarp that comprises an unembayed mountain front. Thick alluvial fans are actively accumulating on the piedmont immediately downstream from the escarpment. This is an example of the type of landscape sketched in Figure 3.

An example of a class 2 terrain in granitic rocks is shown in Figure 5. Although rugged, the cross-valley profile of Wilson Canyon is more U-shaped than V-shaped. An embayment extends upstream from the topographic escarpment that marks the mountain front, and the terraces upstream from the embayment are clear evidence that the floodplain is much narrower than the valley floor. A fault scarp is not apparent and the embayments have created a more sinuous mountain front than probably was formerly present.

The unentrenched alluvial fan downslope from the mountain front shown in Figure 5 should not be considered as evidence for a class 1 front, particularly in view of the types of landscape elements noted above. The unentrenched nature of the fan is the result of the impact of the Pleistocene-Holocene climatic change on a rock type that is sensitive to climatic perturbations (Bull, 1976). Late Pleistocene climates in the Argus range were more conducive to denser plant growth on the hillslopes and for weathering of the hillslope materials than Holocene climates. The postulated decrease in vegetative density during the early Holocene resulted in rapid erosion of the unprotected materials on the slopes. Granitic colluvium responds more rapidly to such a perturbation because lower stresses are needed to move grus-size particles than blocks of rock that would be weathered from most metamorphic rocks. The resulting increase in sediment yield completely backfilled any entrenched stream channels in the valley or on the alluvial fan. In Homewood Canyon, 9 km to the north of Wilson Canyon

130

Figure 5. Class 2 mountain front on the east side of the Argus Range at Wilson Canyon.

131

the same sequence of events has occurred. The stratigraphy exposed in a streambank in the valley of Homewood Canyon consists of well-sorted, reddish-brown, clayey grus of apparent Pleistocene age that is overlain by 2 m of gray bouldery grus of apparent Holocene age. With continuing decreases in the amounts of colluvium left on the hillslopes during the Holocene, the stream systems in both drainage basins have changed their mode of operation and now are downcutting through the alluvium that accumulated on the valley floors.

Mountains that are typical of the class 3 landscape south of the Garlock fault are shown in Figure 6. In contrast to the class 1 setting of the Panamint Range, the topographic highs tend to be occupied by rock types that are resistant to weathering and erosion. Large drainage basins with mountainous areas are hard to define, and the defining of topographic mountain fronts that extend for more than 10 km can be made only on a highly subjective basis. Much of the terrain consists of inselbergs and small mountain masses that are spaced in such a way as to suggest the absence of larger intermontane valleys. Stream systems tend to be discontinuous, and the presence of dune sand on the west sides of some mountains is an additional complication in studying the fluvial systems.

A most useful parameter for assessing the relative tectonic activity of a given mountain front is the thickness of deposits in the basin immediately downslope from an escarpment. Alluvial fans that are thicker than 100 m most likely are the result of a tectonic perturbation, rather than being caused by climatic variations such as were described for the Argus Range. Class 3 terrains may have virtually no basin deposits next to the mountains or may have fans as thick as 10 m that are the result of climatic perturbations. Unfortunately, alluvium thickness data are difficult to obtain, the best sources being from boreholes, entrenched stream channels, and geophysical surveys.

A gravity map of much of the study region (Nilsen and Chapman, 1971) shows anomalies in the basins north of the

Figure 6. Class 3 terrain southeast of the Tiefort Mountains. Soda Mountains are in the background.

133

Garlock fault that are indicative of great thicknesses of
deposits. South of the fault, gravity anomalies appear to
be minor and are chiefly the result of local variations in
the density of surficial rock types. Both the geomorphic
and geophysical evidence suggest that the southern subarea
may be a batholith with low relief that is capped locally
with metasedimentary, volcanic, and Cenozoic terrestrial
rocks. Quaternary faulting is minor or absent in the
southern subarea. One exception appears to the north side
of the Tiefort Mountains (Figure 1). Here a 7 km long,
straight mountain front trends east-west and is associated
with small V-shaped canyons and alluvial fans. This minor
class 1 front may be a product of the stress field associated
with either the Garlock fault, and/or the zone of northwest
trending right-lateral strike-slip faults in the central
Mojave Desert.

The oldest alluvium associated with class 1 fronts
generally is of Holocene age. Late to mid-Pleistocene
piedmont alluvium is associated with most class 2 fronts
and patches of early Pleistocene alluvium in pediment embayments are typical of the class 3 fronts. Any front that
is associated with a granitic drainage basin may have a
piedmont that is dominated by Holocene alluvium, because of
the sensitivity of granitic rocks to climatic variations.

Another useful type of data is the age of the oldest
alluvium downslope from the mountains. Active deposition
of alluvial fans next to the mountains is indicative of
class 1 tectonic conditions, except where such deposition
has been caused by climatic perturbations (Figure 5). The
presence of older alluvium next to the mountains provides
clues as to the length of time that has passed since class
1 conditions last prevailed. Within the arid parts of the
study area, Holocene soils lack the argillic horizons that
are so common on the late and middle Pleistocene geomorphic
surfaces. Early Pleistocene geomorphic surfaces generally
have been dissected into a ridge and ravine topography that
is not conducive to the preservation of soil profiles for
long periods of time.

Areal Variations in Quaternary Tectonism

Pronounced contrasts in the rates and magnitudes of Quaternary uplift are reflected in the landscapes north and south of the Garlock fault. The northern subarea is the southern extent of one of the most tectonically active regions in North America. Class 1 and class 2 mountain fronts are typical of the northern subarea (Table 2) where the Panamint Range rises to more than 3200 m and Death Valley has been depressed to below sea level. In addition to the tilting of the fault blocks, the mean altitude of the northern subarea may be higher than the transitional subarea (significant at $\alpha = 0.20$). Hooke (1972) uses geomorphic criteria to estimate a maximum possible uplift rate for the west front of the Black Mountains at 7 m per 1,000 years, and Quaternary studies by Smith (1975) indicate that the central part of Panamint Valley is the most active tectonically. Fault scarps and faults in the piedmont alluvium also are common in the northern subarea (Hunt and Mabey, 1966; Smith and others, 1968).

Although the Garlock fault marks the southern extent of the prominent north-south structural blocks (see Figure 2) that have been associated with the extensional tectonics of the Basin and Range Province, the magnitudes of Quaternary uplift do not change abruptly at the fault. The north-south mountain fronts north of the Garlock fault have morphologies that reveal decreasing magnitudes of Quaternary uplift toward the south (see Table 2). Within the northern subarea the class 1 fronts change to class 2 fronts, which then change to the class 3 fronts of the transitional subarea. The north-south decrease in uplift rates is most apparent for the Argus Range block, which consists almost entirely of granitic rocks that are susceptible to pedimentation. The fact that the general locations of the mountain fronts are still discernible in the transition subarea, but are extremely poorly defined in the southern subarea is suggestive that the southern subarea has been even more inactive tectonically during the late Cenozoic than the transitional subarea.

Another conclusion is that the general model of the Garlock fault being the result of extensional tectonics to the north of the fault may be too simple. A more complete model is needed to explain the north to south decrease in Quaternary uplift along the mountain fronts north of the fault, and the possibly greater mean altitudes of the northern subarea.

REFERENCES CITED

Bull, W. B., 1973, Local base-level processes in arid fluvial systems: Geol. Soc. America Abstracts with Programs, v. 5, p. 562.

_____, 1976, Sensitivity of fluvial systems in hot deserts to climatic change: American Quat. Assoc. 4th Biennial Conf., discussant paper, p. 42-43.

Cannon, P. J., 1976, Generation of explicit parameters for a quantitative geomorphic study of the Mill Creek Drainage Basin: Oklahoma Geology Notes, v. 36, no. 1, p. 3-16.

Davis, G. A., and Burchfiel, B. C., 1973, Garlock fault; an intracontinental transform structure, southern California: Geol. Soc. America Bull., v. 84, p. 1407-1422.

Davis, W. M., 1909, Geographical Essays, New York, Dover Publications, 777 p.

Denny, C. S., 1965, Alluvial fans in the Death Valley region, California and Nevada: U. S. Geol. Survey Prof. Paper 466, 62 p.

Drewes, Harald, 1963, Geology of the Funeral Peak quadrangle, California: U. S. Geol. Survey Prof. Paper 413, 78 p.

Hamilton, Warren, and Myers, W. B., 1966, Cenozoic tectonics of the western United States: Rev. Geophysics, v. 4, p. 509-549.

Hooke, R. LeB., 1972, Geomorphic evidence for late-Wisconsin and Holocene tectonic deformation, Death Valley, California: Geol. Soc. America Bull., v. 83, p. 2073-2098.

Hunt, C. B., and Mabey, D. R., 1966, Stratigraphy and structure Death Valley, California: U. S. Geol. Survey Prof. Paper 494-A, 162 p.

Jennings, C. W., (compiler), 1958, Death Valley sheet: California Div. Mines and Geology Map Sheet, scale, 1:250,000.

_____, Burnett, J. L., and Troxel, B. W., (compilers), 1962, Trona sheet: California Div. Mines and Geology Map sheet, scale, 1:250,000.

Maxson, J. H., 1950, Physiographic features of the Panamint Range, California: Geol. Soc. America Bull., v. 61, p. 99-114.

Nilsen, T. H., and Chapman, R. H. (compilers), 1971,
    Bouger gravity map of California, Trona Sheet:
    California Div. Mines and Geology Map Sheet, scale,
    1:250,000.

Smith, G. I., 1962, Large lateral displacement on Garlock
    Fault, California, as measured from offset dike swarm:
    Am. Assoc. Petrol. Geologists Bull., v. 46, p. 85-104.

_____, Troxel, B. W., Gray, C. H., Jr., and von Huene,
    R. E., 1968, Geologic reconnaissance of the Slate
    Range, San Bernardino and Inyo Counties, California:
    California Div. Mines and Geology Spec. Rept. 96,
    33 p.

_____, and Ketner, K. B., 1970, Lateral displacement
    on the Garlock fault, southeastern California, sugges-
    ted by offset sections of similar metasedimentary rocks:
    U. S. Geol. Survey Prof. Paper 700-D, p. 1-9.

Smith, R. S. U., 1975, Late-Quaternary pluvial and tectonic
    history of Panamint Valley, Inyo and San Bernardino
    Counties, California: California Institue of Techno--
    logy Unpub. Ph.D. Dissertation, 295 p.

Wright, L. A., and Troxel, B. W., 1967, Limitations on
    right-lateral strike-slip displacement, Death Valley
    and Furnace Creek fault zones, California: Geol. Soc.
    America Bull., v. 78, p. 933-950.

A TENTATIVE SEDIMENT BUDGET FOR
AN EXTREMELY ARID WATERSHED
IN THE SOUTHERN NEGEV

Asher P. Schick

Department of Geography
The Hebrew University of Jerusalem

ABSTRACT

A sediment budget for the 0.5 km^2 watershed of Nahal
Yael is attempted on the basis of detailed observations of
rainfall, streamflow, sediment transport and deposition, and
changes in geomorphic features. For the ten year period
considered, the mean annual rainfall was 31.6 mm , 99 percent
of the geomorphic work was accomplished during 5 days by 7
discrete events.

Sediment in suspension, computed from data obtained
through an automatic sampling program, accounted for a mean
annual yield of 127 tons. Bedload yield is estimated on
the basis of, (1) distance of transport determined from
traceable particles, (2) the area and depth of the scour
layer for the inner channel bed and for the gravel bars, and
(3) a comparison of grain sizes on the bed and the bars with
sediment in transport as sampled by liquid samplers and
bedload traps. A mean annual yield of 66 tons was found.
The dissolved load is about one percent of the total load.
The resulting mean annual sediment yield of 388 tons/km^2
considerably exceeds the accepted norm for arid environments.
It also exceeds by a factor of 3 the estimated sediment
yield, corrected for drainage area, for the 3,100 km^2 watershed of Wadi Watir, located in a similar environment in
eastern Sinai.

While the internal sediment delivery ratios of the
Nahal Yael drainage system are reasonably consistent, the
aggradation rates as measured directly on the alluvial fan
over a period of six years are one fifth of what they should
be on the basis of the sediment transport computation.

Although inadequacies of the sampling and measurement program may explain a part of the discrepancy, the main reasons
are an insufficient understanding of the transport mechanism
during violent desert floods, exchanges between suspended
and bedload transport modes, and the importance of transient
alluvial storage and its intense localization.

INTRODUCTION

The purpose of this paper is to construct a sediment
budget for a small arid watershed and to use the quantitative
relationships of this balance as a vehicle for describing
and enhancing the understanding of the dominant geomorphic
processes in extreme deserts.

Only few studies provide us with a fairly complete
watershed sediment budget. Among these, the best known ones
represent a large watershed in a mid-latitude highland area
(Jaeckli, 1957), a small watershed in a high-latitude mountain area (Rapp, 1960), and a small watershed in a semi-arid
hilly area (Leopold and others, 1966). All these studies encountered difficulties in sustaining the balance by equaliz-
ing denudation from upstream source areas with larger watersheds of which they form a part. The downstream diminution
of denudation per unit area, long ago recognized by engineers
under the term "sediment delivery ratio", is in reality a
problem of magnitude-frequency relationships which affect
the alluvial storage (Trimble, 1976).

A quantitative process study in an extremely arid environment cannot hope to duplicate the standards of accuracy
of similar studies in more benign climatic zones. The important geomorphic events are rare, violent, and unpredictable,
while logistics is very complex. There are also some minor
advantages. The lack of soil, vegetation, and interflow sim-
plify the water and erosion mechanism. Population sparseness
results in less vandalism. Events are highly discrete, with
long intervening periods of geomorphic stagnation which enable undisturbed documentation.

On the whole, however, sediment budget studies in the
desert must be satisfied with less direct and dependable

measurements and with more assumptions than would be the
norm in more humid environments. The rarity of the data
is the main reason for the presentation here of such a
budget.

NAHAL YAEL

The venue of this study is the Nahal Yael watershed in
the southern Negev, Israel. The site as well as its
characteristics in terms of physiography, hydrometeorology,
slope and fluvial processes have been described in some
detail (Gerson and Inbar, 1974; p. 16-18; Schick, 1970, 1974,
Schick and Sharon, 1974; Sharon, 1970; Yair and Klein, 1973).
Briefly, Nahal Yael is a 2 km long ephemeral stream draining
0.6 km^2 of bare rocky desert into the major wadi of Nahal
Roded, which debouches into the Arava rift valley near the
northern end of the Gulf of Aqaba (Figure 1). Midday summer
temperatures approach 50° C. At nearby Eilat, mean annual
rainfall barely exceeds 25 mm.

The watershed is rugged, with slope angles in many
parts exceeding 25° (Figure 2). Fluvial dissection is
well developed. Over 100 fingertip tributaries drain the
upland areas toward the 1.0 km long main alluvial valley.
The overall longitudinal slope of the alluvial channel
is 0.05. Further downstream the channel widens to form an
alluvial fan whose toe borders on the wide braided channel
of Nahal Roded (Figure 3).

Geologically the Nahal Yael watershed is part of the
northern extension of the Sinai Massif. In detail it is
composed of a fine and complex pattern of schists (one
half of the watershed, in its central part), various basic
rocks, granite, and numerous dikes of variable nature.
Some slopes, especially in upland areas, are relatively
smooth, but most are covered by a 5-20 cm thick rubbly
layer of angular or platy stones 5-50 cm in size. Material
of sizes which later turn up as suspended load in floodwaters is not visibly exposed on most of the surface.

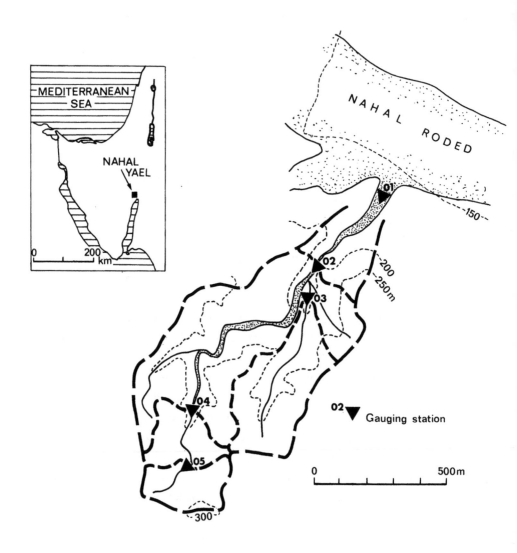

Figure 1. Location map, Nahal Yael, Southern Negev, Israel. Detailed map shows location of gauging stations and the pertinent watersheds.

A system of 1-5 m high terraces parallels the main alluvial reach for most of its length. There is no vegetation in the watershed area, with the exception of several acacias and small bushes which grow in channel bed reaches locally favored by sub-alluvial water storage.

Figure 2. Lower part of main alluvial reach, Nahal Yael. View upstream. Gauging station 02 in left foreground. Main alluvial reach and valley rise to the right; tributary 03 -- to the left.

143

The observation program at Nahal Yael, which started in 1965, used a variety of measurement systems whose number and location were changed according to needs. The stream gauging network was, however, kept unchanged in location and number of stations, since its inception in 1966. These stations were determined by geomorphic considerations. Stations 05, 04, and 03 (Figure 1) gauge small upland watersheds with little routing and negligible alluvial storage. Station 04 is located at the transition between the bedrock channel upstream and the reach of continuous alluvium; later also termed "main reach". The downstream end of this main reach is a 10 meter high dry waterfall in granite, probably caused by erosive events in the Miocene. Station 02 is located on the bedrock lip of this watershed. As a result, a part of the alluvial underflow in the main alluvial reach is forced back on the channel surface. The apex of the Nahal Yael alluvial fan starts 100 m downstream of the waterfall. Station 01 is located on the main channel outlet at the toe of the fan.

WATER BALANCE

During the ten year period considered, mean annual rainfall was 31.6 mm (Table 1). Extremes of annual rainfall ranged from 1.9 to 66.9 mm. Over 20 percent of this rainfall, again for the ten year period, became runoff inside the watershed, but ultimately less than 5 percent flowed into Nahal Roded (Table 2).

The geomorphically significant runoff was caused by seven major events. The longest period between subsequent major events was over 45 months, and the two shortest periods were 4 hours (event 7A and 7B, on 21 January 1973) and 8 hours (event 12A and 12B, on 21 February 1975).

A generalized water balance (Figure 4) shows that watershed 02 contributes 20 runoff units to the alluvial fan, but only 4 units exit to Nahal Roded as surface flow. Sixteen units are lost by infiltration on the fan surface.

Figure 3. The alluvial fan of Nahal Yael and its junction with Nahal Roded. Photo by W. B. Bull.

Water Year	Annual rainfall, mm	Remarks
1966/67	13.2	Eilat Meteorological Station
1967/68	66.9	Eilat Meteorological Station
1968/69	45.4	Recorder 25
1969/70	1.9	Recorder 25
1970/71	29.6	Recorder 25
1971/72	26.5	Recorder 25
1972/73	11.8	Recorder 25
1973/74	46.1	Recorder 25
1974/75	64.0	Recorder 25
1975/76	10.9	Recorder 26

Mean, 1966/67 - 1973/74 (8 years):   30.2 mm

Mean, 1966/67 - 1975/76 (10 years):   31.6 mm

Table 1. Annual rainfall, Nahal Yael, 1966/67 - 1975/76.

145

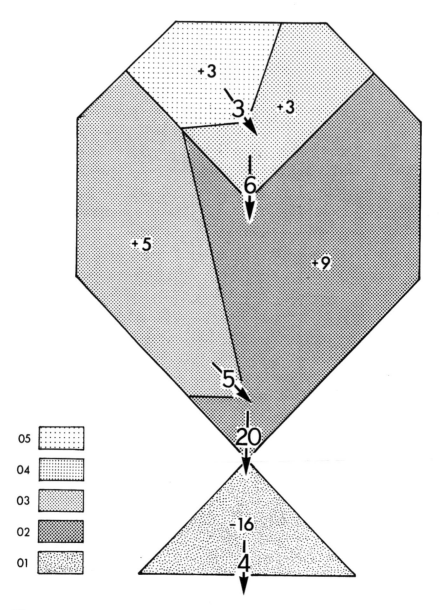

Figure 4. *Generalized water balance, Nahal Yael, 1966/67 - 1975/76.*
One unit equals one percent of total ten year rainfall on
entire watershed (01).

Watershed	01	02	03	04	05
Area (km^2)	.58	.50	.13	.11	.05

Eight year period 1966/67 - 1973/74

Total rainfall (m^3)	140,000	121,000	31,400	26,600	12,100
Total runoff (m^3)	3,190	17,300	5,050	6,840	3,130
Rainfall-runoff ratio %	2.3	14	16	26	26

Ten year period 1966/67 - 1975/76

Total rainfall (m^3)	183,000	158,000	41,000	34,800	15,800
Total runoff (m^3)	8,090	36,400	10,000	11,800	5,730
Rainfall-runoff ratio %	4.4	23	24	34	36

The second part of the table includes event 12 whose frequency probably exceeds 10 years.

Table 2. *Ten year rainfall and runoff volumes, Nahal Yael, 1966/67 - 1975/76.*

In the upper part, watersheds 05, 04, and 03 contribute together 11 units of runoff, while the complementary area of 02, which is slightly more than one half of the total area of 02, contributes only 9 units. Assuming that the small tributaries and valley sides, which are symmetrical to the main reach, have a similar response in both areas, we expect about 12 units of runoff to be supplied to the main alluvial reach. Hence, it is reasonable to assume that 3 runoff units are lost by infiltration through the main alluvial reach between stations 04 and 02.

Most of the runoff events are highly peaked and short-lived (Schick, 1971). Peak discharges for the small upstream watersheds exceeded several times 10 m^3/sec/km^2 (Figure 5). Probably event 12B, which caused the highest flood recorded at any of the watershed stations, exceeds in frequency the ten year period of observation. The exclusion

147

of this event from the long-term rainfall and runoff considerations by computing eight year values affects the mean
annual rainfall only slightly (Table 1), but causes a
significant decrease in rainfall-runoff ratios in all
watersheds. The basic inflow-outflow mechanism of water
as represented for the ten year values by Figure 4 remains,
however, unchanged.

SEDIMENT IN SUSPENSION

Over 300 samples of suspended sediment were collected
during the study period in the watershed and in similar but
larger streams in the vicinity. Most of these came from
simple automatic instruments which operate only on the rise,
but some come from manual sampling and from a time-interval
operated instrument, both of which concentrated on the
falling stage.

Mean concentration values of suspended sediment show
a consistent relationship with peak flow and frequency of
event (Figure 6). The geomorphic location of the sampling
station and thus of the contributing watershed is of prime
importance. For a normal major event, frequency between
2.5 and 5 years, the mean concentration of suspended sediment in the floodwaters from the fan toe is four times
higher than from the upland tributaries.

Table 3 shows the estimated yield of suspended sediment
for each event and watershed of Nahal Yael. Several events
show a better internal consistency than others. Some of
the differences may well be due to inaccuracies inherent
in the method. However, the general trend is clear, after
correction for area, watershed 02 yields somewhat higher
overall quantities of suspended sediment than its component
watersheds 03 and 04 seem to indicate. Compared to a mean
annual inflow of suspended sediment into the main alluvial
reach of about 80 tons, there is an outflow toward the fan
of 127 tons. Thus, a systematic degradation of the main
alluvial reach is indicated.

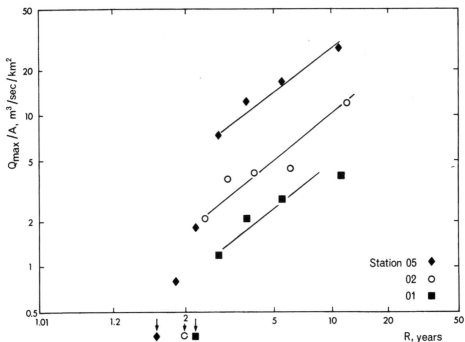

Figure 5. Annual floods, Nahal Yael, stations 01, 02, and 05. Drainage areas .58, .50, and .05 km², respectively. Period of observation: for 02 - 11 years (October 1965 to September 1976); for 01 and 05 - 10 years (October 1966 to September 1976). Q_{max}/A = annual peak flow per unit drainage area, in cubic meters per second per square kilometer; R = Recurrence interval in years.

Station 01 at the fan toe, which was significantly active only four times during the ten year period, shows an outflow only one half as high as its sediment inflow through station 02, near the fan apex. The probable reason for this discrepancy is the systematic aggradation of the fan during the period considered.

The median size of the particles which constitute the inner channel bed of the main alluvial reach is 2-3 times larger than the material sampled in transit (Figure 7). An assumption stating that the finer half of the sediment in transit as sampled derives from diffuse slope wash (Yair and Klein, 1973) would bring the coarser half in close agreement with the inner channel bed material. Allowance must also be made for the difficulty of coarse sand grains and granules to enter the 9 to 11 mm wide intakes of the sampling instruments at the reduced filling velocities.

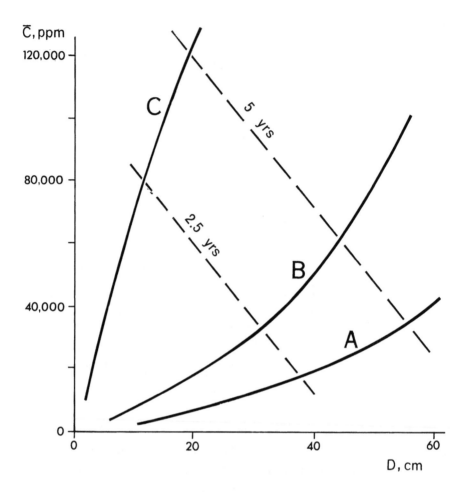

Figure 6. *Probable mean concentration of suspended sediment $\bar{C}$ in rela-*
tion to depth of peak flow D and to frequency of event for
various geomorphic locations.
A - upstream part of watershed, bedrock channel
B - mid-part of watershed, alluvial channel
C - toe of alluvial fan at watershed outlet

BEDLOAD

The evaluation of bedload discharge presents one of
the most difficult problems in geomorphology. Within
the framework of approximations used in this sediment
budget, the method used was very simplistic. First,

Event	Date	01	02	03	04	05
3	4-25-68	3.2	82	7.5	3.0	1.0
4	5-24-68	77	108	23	18	6.6
5	11-24-68	0	0	0	0.01	0.01
6	11-25-68	0	0.1	0.5	1.3	0.3
7A	1-21-69	0	55	3.1	1.8	3.3
7B	1-21-69	86	131	15	7.1	3.2
8	3-25-71	0	0	0	1.7	0.2
9	12-21-71	0	0	0	6.6	0.5
10	11-12-73	90	204	59	65	33
11	12- 4-74	0	11.2	1.0	5.9	7.6
12A	2-21-75	0	200	27.1	14.9	2.0
12B	2-21-75	375	479	73.5	47.6	23.5

Total, 8 yrs. (1966/67-1973/74)

		256	580	108	105	48

Total, 2 yrs. (1973/74-1975/76)

		375	690	102	68	33

Total, 10 yrs. (1966/67-1975/76)

		631	1270	210	173	81

Mean annual yield of suspended sediment

		63.1	127	21.0	17.3	8.1

Table 3. Yield of suspended sediment, Nahal Yael watersheds, 1966/67 - 1975/76, in tons.

the depth of scour was determined on the basis of a few data from several chains in Nahal Yael and in comparison with other studies such as Leopold and others (1966). Second, the transport distance of the scour layer downstream was approximated on the basis of extensive data on the movement of labelled particles through the main alluvial reach of Nahal Yael. Third, a two-phase approach,

151

wherein, the inner channel bed and the gravel bars were treated separately, enabled estimates of bedload discharge per "mean" major event and hence, the derivation of a mean annual value.

The median particle size of bedload trapped in several simple bedload traps located in the vicinity of station 02 of Nahal Yael exceeds by a factor of 2 or more the median size of the particles which constitute the shallower in-channel bars (Figure 8). The trap has a wire netting with openings of 35-50 mm. This enables a part of the finer bedload to escape. Also, the in-channel bars considered may include some surficial admixture of in-channel bed material. Considering both factors, it may be assumed that the coarse bedload in transit is indeed the gravel bar material.

The time-weighted depth of the scour layer, estimated as a flat 40 percent of the "mean" maximum water depth for a major event, is 8.6 cm for the inner channel and 3.8 cm for the bars (Table 4). The area of the inner channel, within the main alluvial reach, is 2,740 m^2 and that of the aggregate area of the gravel bars (i.e., the probable flood channel less the inner channel) is 12,500 m^2 (Table 5). Data on transport distance of particles show a mean value per event of 65 m, with little change for varying particle size (Table 6). There is a tendency for particles smaller than 32 mm to move only 40-45 meters per event, but this is disregarded because dilution, burial, and loss of identification affect the smaller particles more than the larger ones.

The bedload discharge is computed in Table 7. It is, for the ten year period considered, 66 tons per year, of which one third is of finer particles, roughly equivalent in size to the inner channel material. It is possible that up to one half of the material entrained from this source actually travels as suspended load due to the turbulent character of the floodwaters. A correction in this direction would lower the annual bedload yield up to

152

(i)   FLOW DEPTH

    Mean maximum depth of flow as registered at
       sediment samples 73, 75, 76 for the seven
       major events   34 cm

    Estimate of "dominant" maximum depth of flow
       based on a survey of flood marks (Figure 9).   42 cm

    "Mean" maximum water depth   <u>38 cm</u>

(ii) SCOUR LAYER, INNER CHANNEL

    "Mean" maximum water depth for reach   38 cm

    "Mean" maximum depth of scour layer, estimated
       as 40 percent of maximum water depth   15.2 cm

    above, corrected for entire duration of flow
       by a reduction of one half (i.e. time-
       weighted)   <u>8.6 cm</u>

(iii) SCOUR LAYER, GRAVEL BARS

    "Mean" maximum water depth over shallow bars,
       estimated as being 9.5 cm above inner
       channel bed   28.5 cm

    "Mean" maximum depth of scour layer on shallow
       bars, estimated as 40 percent of maximum
       water depth   11.4 cm

    above, corrected for entire duration of flow
       by a reduction to one third (i.e. time-
       weighted)   <u>3.8 cm</u>

Table 4.   *Computation of depths of flow and estimates of depth of scour
layer, main alluvial reach, Nahal Yael, 1966/67 - 1975/76,
in centimeters.*

11 tons, and is therefore insignificant at the level of
accuracy employed here.

    The bedload yields determined apply strictly to station
02 only, as they were computed by utilizing the geometry
and other characteristics of the main alluvial reach of
Nahal Yael at whose downstream end this station is located.

	Inner channel	Probable flood channel
Main channel, DM 3940 to 4920	2,250	14,100
Major tributary, length of continuous alluvium 150 m, approximated by DM 3940 to 4090	244	650
Four minor tributaries, mean length of continuous alluvium 40 m, approximated by DM 3940 to 3980	246	480
Total, probable flood channel area		15,200 m²
Total, inner channel area regarded as main source of finer bedload	2,740 m²	2,740
Total, gravel bar portion of probable flood channel area, regarded as main source of coarser bedload		12,500 m²

Table 5. *Computation of inner channel and probable flood channel areas, Nahal Yael, in square meters. Based on Figure 7.*

SEDIMENT YIELD AND STORAGE

The total sediment yield for Nahal Yael at station 02 is derived by summing up the yields of suspended sediment, bedload, and sediment in solution as follows:

Suspended sediment	127 tons
Bedload	66 tons
Sediment in solution	1.2 tons
Total mean annual sediment yield	194 tons = 388 tons/km²

The dissolved content of the Nahal Yael floodwaters was found to be on the order of 300 mg/l. This concentration equals an annual yield of 1-2 tons; less than one percent of the mechanical components of the sediment.

Thus we have, for 02, a total mean annual sediment yield of about 400 tons per square kilometer of drainage area, with roughly one third coming as bedload.

The alluvial fan, which begins a very short distance downstream from station 02, is about 200 m long and has a maximum active width of 120 m. The active part is delimited to the east by a 5-20 m high scarp of Pleistocene clastic material deposited by the Nahal Roded system. Along its western boundary a 0.5-3 m high scarp rises to an old and presently inactive fan of Nahal Yael.

Station 01, at the toe of the fan and at the confluence with Nahal Yael with Nahal Roded, can be regarded as measuring the water and sediment outflow from the fan. Practically all the inflow to the fan comes from station 02.

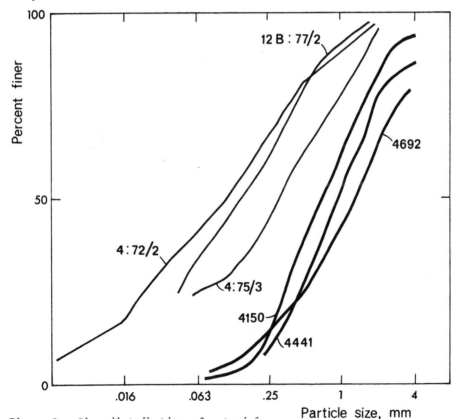

Figure 7. Size distribution of material
(1) on inner channel bed (at DM 4150, 4441, 4692);
(2) from floodwater samples (4:72/2, 4:75/3, 12B:77/2).

Particle generation	Date introduced	Number of major events transported	Number of particles	Median size, mm	Mean distance transported per event, m
Old red	4 - 1-65	3*	1,411	45	63-109
Orange	2-11-68	4	101**	8	40
		4	111**	28	45
		2	56	55	51
New red	9 - 9-68	2	226	84	22
		2	28**	121	14
Blue	9 - 9-71	1	234	85	90
		1	24**	124	36

Mean transport distance per event for size class of 20-80 mm   <u>65 m</u>

* Survey following fourth flood useless because of dilution and discoloration

** Number indicates total of retrieved particles

Table 6.   Transport distances of bedload particles, Nahal Yael.

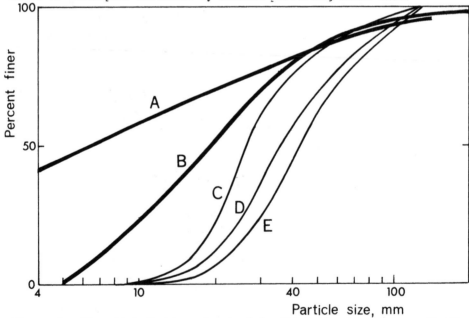

Figure 8.   Size distribution of material (1) on in-channel bars (A, B); (2) collected in bedload traps (C, D, E).

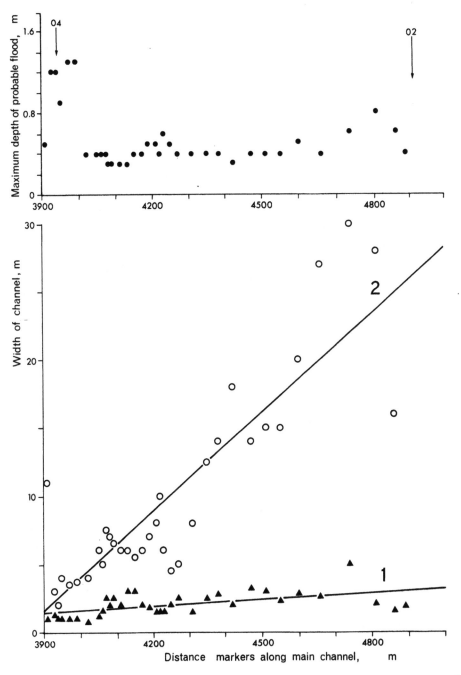

Figure 9. *Downstream change: in depth of "dominant" flood (D_{max}), based on a survey of flood markers, and in width of (1) inner channel and (2) probable flood channel, main alluvial reach, Nahal Yael. Gauging stations 04 (at DM 3940) and 02 (at DM 4920) are located at the upstream and downstream ends, respectively, of the continuous mid-basin alluvial valley of the watershed.*

	Coarser particles	Finer particles	Total
Area contributing, m^2	12,500	2,740	
Mean depth of scour, m	0.038	0.086	
Mean volume participating in movement, m^3	475	236	
Mean distance of transport per event, m	--- 65 ---		
Length of alluvial reach, m	--- 980 ---		
Number of major events needed to yield entire volume	--- 15 ---		
Bedload yield per "mean" major event, m^3	32	16	48
Number of major events in ten year period (1966/67-1975/76)	--- 7 ---		
Bedload yield for ten year period, m^3	220	110	330
Estimated mean annual bedload yield, (tons/yr.)	44	22	66

Table 7. Computation of bedload discharge at downstream end of main alluvial reach (station 02), Nahal Yael.

About one half of the difference in area between watersheds 02 and 01 (.08 km^2) is the area of the fan itself and the rest is mostly short slopes on granite and on terrace material. Large scale losses of water by percolation into the fan are documented by the data (Figure 4), but for sediment the fan must act as a system wherein, the difference between input at 02 should equal output at 01 less changes of volume of the fan.

Seven sections which run across the fan have been re-surveyed at regular intervals since 1967. For the purpose of this summary, the net aggradation accomplished by events 7A, 7B, and 10 is considered (Table 8). The total lag of material left on the fan, presumably due to massive infiltration, is still five times higher than

158

the aggraded material as determined through the section, even if only bedload is considered.

Net aggradation, alluvial fan, 27 April 1967 to 14 November 1973, as determined by cross-section surveys		10.57 m^3	21 tons
Sediment inflow, station 02, same period (events 7A, 7B, 10):			
(i) suspended		390 tons	
(ii) bedload ($\frac{1}{2}$ added)		195 tons	585 tons
Sediment outflow, station 01, as above:			
(i) suspended		176 tons	
(ii) bedload ($\frac{1}{2}$ added)		88 tons	264 tons
Deduced aggradation on fan,			
total load			321 tons
bedload only			107 tons

Table 8. Tentative sediment budget for the Nahal Yael alluvial fan.

DISCUSSION

In watershed sediment budgets, inaccuracies due to sampling problems, areal and time generalizations, and other factors are inherent and substantial. Probably, they are larger in this study than in most others. At the same time, major discrepancies have been reported more often than might be caused by inaccuracies alone. To quote but a few studies, in a semi-arid area in western Colorado, only one-third of the amount of sediment yield from areas of 0.5 to 5.2 km^2 is transported through the system (Hadley and Shown, 1976). In the Arroyo de los Frijoles, New Mexico, also a semi-arid area, sheeterosion, estimated basin-wide from erosion pins, gives a value of sediment production 45 times larger than can be accounted for by channel aggradation and reservoir accumulation. Similar findings come from the Cheyenne River basin (Hadley and Schumm, 1961). A long-term study based on numerous scour chains at a large number of sites in western U. S. confirms the ineffectiveness of the sediment conveyance system by events of "normal" frequency (Emmett, 1974).

Changes of phase between bedload and sediment in suspension is probably important in most events. In addition, some of the material entrained on the inner channel and which figures in the computation as fine bedload, may actually travel and be sampled as suspended particles. However, both these complications cannot change the resulting total yield by more than a factor of two. Thus, the major budgetary discrepancy of the alluvial fan remains unexplained.

Another line of possible explanation involves localization of deposits on the fan. To monitor the aggradation-degradation changes on the 7,000+ m^2 fan area, 300 m total active length of sections were used. It is possible, though not probably, that massive infiltration in zones located between, rather than on, surveying section lines, remain undetected by the volumetric procedure. Visible field evidence which would support such a hypothesis, such as sedimentary structures, has not been found, but intense localization of transient alluvial storage cannot be entirely disregarded as a possible partial explanation.

Import or export of material by wind can be another complication in watershed sediment budgeting. Although wind erodibility in the Negev is potentially high (Yaalon and Ganor, 1966), only the finer part of the load equivalent to our suspended sediment can be involved in deflation off the fan. Thus, again, a correction factor of more than 50% of the suspended sediment yield is highly unlikely.

The sediment transport mechanism during violent desert floods, especially those with "walls of water" and supercritical flow velocities, is still imperfectly understood. Further observations and process studies are needed to overcome this gap in knowledge, which is probably the main reason for our inability to balance sediment budgets in arid watersheds.

ACKNOWLEDGMENTS

The research reported here has been carried out with the support of the European Research Office of the U. S. Army, London, and of the Central Research Fund, The Hebrew University of Jerusalem. Judith Lekach assisted with the computation. The figures were drawn by Tamar Sofer.

REFERENCES CITED

Emmett, W.W., 1974, Channel aggradation in western United States as indicated by observations at vigil network sites: Zeitschr. fur Geomorph., N.F., Suppl. Bd. 21, p. 52-62.

Gerson, R. and Inbar, M., 1974, The field study program of the Jerusalem-Elat Symposium, 1974; reviews and summaries of Israeli research projects: Zeitschr. fur Geomorph., N.F., Suppl. Bd. 20, p. 1-40.

Hadley, R.F. and Schumm, S.A., 1961, Hydrology of the upper Cheyenne River Basin: U.S. Geol. Survey Water Supply Paper 1531-B, 198p.

_____ and Shown, L.M., 1976, Relation of erosion to sediment yield: Proc., 3rd Federal Inter-Agency Sedimentation Conf., Denver, Colo., 1976, Symp. 1, p. 132-139.

Jaeckli, H., 1957, Gegenwartsgeologie des buendnerischen Rheingebietes: Beitraege zur Geologie der Schweiz, Geotechnische Serie Lieferung 36, 136p.

Leopold, L.B., Emmett, W.W. and Myrick, R.M., 1966, Channel and hillslope processes in a semi-arid area, New Mexico: U.S. Geol. Survey Prof. Paper 352-G, p. 193-253.

Rapp, A., 1960, Recent developments of mountain slope in Karkevagge and surroundings, northern Scandinavia: Geog. Ann., v. 42, p. 71-206.

Schick, A.P., 1971, Desert floods - interim results of observations in the Nahal Yael research watershed, Southern Israel, 1965-1970: Internat. Assoc. Hydr. Sci., Publ. 96, p. 478-493.

_____, 1974, Formation and obliteration of desert stream terraces -- a conceptual analysis: Zeitschr. fur Geomorph., N.F., Suppl. Bd. 21, p. 88-105.

_____, and Sharon, D., 1974, Geomorphology and climatology of arid watersheds: Mimeogr. Report, Dept. of Geography, Hebrew Univ., Jerusalem, 161p.

Sharon, D., 1970, Topography- conditioned variations in rainfall as related to the runoff-contributing areas in a small watershed: Israel Jour. Earth Sci., v. 19, p. 85-89.

Trimble, S.W., 1975, Denudation studies: can we assume stream steady state?: Science v. 188, p. 1207-1208.

Yaalon, D.H. and Ganor, E., 1966, The climatic factor of wind erodibility and dust blowing in Israel: Israel Jour. Earth Sci., v. 15, p. 27-32.

Yair, A. and Klein, M., 1973, The influence of surface properties on flow and erosion processes on debris covered slopes in an arid area: Catena, v. 1, p. 1-18.

SEDIMENT ORIGIN AND SEDIMENT LOAD
IN A SEMI-ARID DRAINAGE BASIN

Ian A. Campbell
Department of Geography
University of Alberta

ABSTRACT

Sediment origin and suspended sediment load data are examined for the large (43,000 km^2) drainage basin of the Red Deer River in southcentral Alberta. Most of the basin lies within the semi-arid prairie environment though the headwaters are in the moist foothill and mountain section of the province. Fifty percent of the river's mean annual discharge of 70 m^3/s are derived from the headwaters region, or about six percent of the total basin area. The remaining runoff is derived from the prairies. Sediment supply from the headwaters region is comparatively small, accounting for perhaps 12 percent of the total suspended sediment load, and much of the prairie region contributes neither runoff nor sediment to the river. Detailed observations of erosion rates (4.50 mm/yr) and estimates of sediment yield (8,217 tonnes/km^2/yr from a region of badlands, that form about two percent of the basin's area, shows these supply about 90 percent of the total suspended load. Parallels are drawn between this situation and the partial area concept of runoff. It is concluded that the common practice of determining regional erosion rates is meaningless.

INTRODUCTION

While there exists a voluminous and ever-growing body of literature dealing with the problems of sediment transport in streams, comparatively little attention has been given to identifying precisely the areas from which the sediment is initially derived and the rates at which it is produced and moved. The standard technique employed in most studies of sediment yields is simply to mathematically re-distribute the stream's sediment load back onto the water-shed in a uniform manner and call it a 'regional erosion rate.' This is despite an increasing amount of evidence to show that the sediment carried by most streams is derived from a relatively minor proportion of the basin area; typi-cally the channel banks and, by extension, the valley sides. Major areas of the basin may be non-contributing as far as sediment is concerned. The concept of a 'regional erosion rate' in these terms becomes essentially valueless though it may be applicable under some highly specialized condi-tions.

The reasons why identification of sediment source areas and their rates of erosion has been relatively neglected as an area for investigation are not hard to find. Erosion on many surfaces is infinitesimally slow, making accurate measurements difficult and tedious. Furthermore, in an environment composed of a multitude of differing surface and vegetative characteristics, the problem of devising an adequate sampling program may prove intractable.

One solution is to seek more 'simple' landscapes in which to work; hence, many investigators have turned to those areas where logistical and temporal problems are reduced, where erosion is rapid and it is compara-tively easy to take measurements and estimate, reasonably accurately, factors such as sediment yield. In this respect, on the larger scale, the arid and the semi-arid lands form nearly ideal work areas since one of the most complicating of the environmental factors determining erosion (namely variations in vegetation cover) is absent or nearly so. It is therefore, no accident that many of the fundamental observations and theories current in geomorphology have been developed from study in the world's drylands. Observations of processes operating in such environments may well be hampered, however, by

the sporadic and unreliable nature of one of the most
powerful of the process-generating agents, precipitation.

The solution to this problem, as seen by several investigators, appears in the form of badlands. Here the
whole range of arid and semi-arid topography is modelled in
miniature, and erosion is unbelievably rapid in terms of
normal geologic rates. Moreover, the general sterility
of the lithological units that favor badland development
so retards vegetation growth that often quite spectacular
badlands develop under climatic conditions which normally
would favor a well-established plant cover. Examples here
would include those 'badlands' associated with geotechnic
activities such as strip-mining and spoil heap formation.
Badlands have provided an ideal outdoor laboratory for
process-oriented geomorphologists. Erosion is rapid, forms
develop quickly and enormous quantities of sediment are
produced from comparatively small areas.

Where badlands occur in drainage basins they invite
attention both in their own right and also because their
potential to generate large sediment yields is so dis-
proportionately great that they provide significant clues
in any attempt to understand the problems associated with
sediment origin and sediment load.

One such basin where badlands exert such a powerful
control on sediment production is the Red Deer Basin in
semi-arid, southcentral Alberta.

THE RED DEER BASIN

The 43,000 km^2 Red Deer Basin lies almost entirely
in the semi-arid eastern Alberta plains where mean annual
precipitation is generally less than 450 mm (Figure 1).
About seventy percent of this precipitation falls in
the four summer months, usually in the form of high-
intensity convective thunderstorms. Only in the extreme
western portion of the basin, in the foothills and front
ranges of the Rockies, is precipitation heavy (>500 mm) and
relatively reliable. The net effect of this climatic

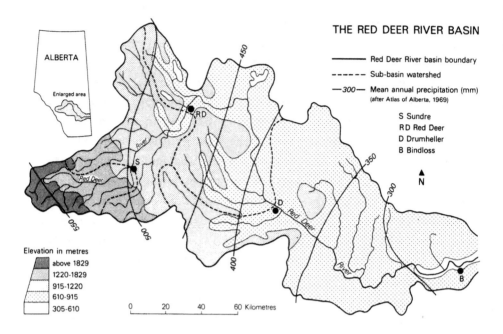

Figure 1. Physical characteristics of the Red Deer River Basin.

pattern is to create disporportionately high amounts of
runoff per unit area on the western side of the basin and
very low runoff amounts in the eastern portions. Figure
2 shows the areal runoff contributions for each sub-unit
of the basin above its respective stream-gauging station.
At Sundre, which drains less than 6 percent of the basin,
the mean annual discharge is about half of that measured
at the lowest downstream station at Bindloss. Conversely,
the sub-basin above Bindloss, almost half the total basin
area, contributes less than 15 percent of the mean annual
discharge. Table 1 shows the set of relationships in more
detail. The dramatic decrease in runoff per unit area
is also shown on Figure 2, again predominantly reflecting
the climatic gradient.

On Figure 3 are plotted the mean annual discharge
values for the Red Deer at the Red Deer gauging station
and those at Bindloss for the period 1967-1973. The
shaded area represents the incoming discharge contributed between these two stations. Within broad limits

Gauging Station	Mean Annual Discharge (m³/s)[1]	Incremental Measure for Sub-basin (m³/s)	Mean Annual Discharge as a Percentage of the total[2]	Contributing Area (km²)[3]	Contributing Area as a Percentage of the Total[4]	Mean Annual Runoff (mm)
Sundre	34.27	34.27	49.60	2,442	5.60	440
Red Deer	51.55	17.28	25.00	8,873	20.60	61
Drumheller	60.55	8.50	12.30	13,414	31.20	19
Bindloss	69.12	9.07	13.10	18,271	42.50	15
Total/Mean	69.12	69.12	100.00	43,000	99.90	50

1   As measured at the station (Alberta Research Council 1972)

2   Calculated as the amount that each station gauges as a percentage of the total measured at Bindloss

3   The area as determined by the outline of the sub-basin between successive gauging stations, or, in the case of Sundre, above Sundre to the outlines of the watershed

4   Calculated as the amount of the area above each station as defined in (3) expressed as a percentage of the total.

Table 1.   Mean Annual Discharge and Contributory areas, Red Deer River, Alberta.

the percentage contribution remains relatively constant
though certain discrepancies do occur, i.e., the discharge
for 1970 and 1971 was almost identical at Red Deer but
1971 was a comparatively high runoff year as measured
at Bindloss. What should be stressed is the fact that
the shaded area represents the contribution from about
70 percent of the entire basin area whereas, the area
beneath the Red Deer graph line shows that derived from
only about 30 percent of the basin (Figure 2).

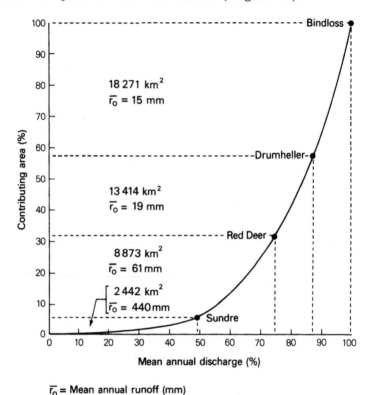

Figure 2. *Contributing areas, runoff and discharges within the Red Deer
Basin.*

This pattern of runoff is typical for many arid and
semi-arid basins and would be expected from a purely climatic viewpoint. It also indicates, however, the effects
of topographic control, since, on the flat or gently
rolling plains there are large areas of interior drainage
that contribute no runoff to the Red Deer. The extent

of these areas has not been accurately determined but it
is probably of the order of 20 to 30 percent of the surface
within the main Red Deer watershed.   There are in addition
substantially higher losses of potential runoff due to evaporation and infiltration in the dominantly flat and treeless
landscape.

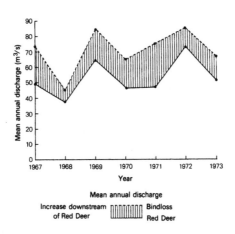

Figure 3.   Mean annual discharges for Red Deer and Bindloss.

The course of the Red Deer, its valley form and tributary system, is geologically recent.   The present river
system is entirely post-Wisconsin in age, having developed
during and since deglaciation about 10,000 to 14,000 years
B.P.   Erosion has been rapid.   The main valley now is
entrenched about 100 m into the plain's surface, giving
a mean rate of downcutting of about 1 m/100 yr.   The fast
erosion was assisted by the weak nature of the underlying
bedrock, which consists almost entirely of Upper Cretaceous
shales and poorly cemented sandstones (Green, 1970).   The
glacial deposits are comparatively thin, rarely exceeding
35 m in thickness and usually much less, and they too
are easily eroded.

Rapid incision by the river produced steep valley
sides whose average slope is about 30-35° and this,
together with the regional geology and the sparse vegetation

cover under a semi-arid climate, promoted the development of badlands.

The badlands are found along both sides of the Red Deer as a more or less, continual, linear feature stretching from about 40 km downstream of Red Deer City to about 20 km upstream of Bindloss (Stelck, 1967). In total, the badlands occupy an area of about 800 km^2, which is only 2 percent of the total basin area. Their sediment production rate overshadows by far, however, their areal significance since they contribute almost all of the sediment carried by the Red Deer River.

The data plotted on Figure 4 show the overwhelming importance of the badlands' sediment yield. The upper line shows the total suspended load values for Bindloss for the period 1967-1974. It is immediately evident that great annual variations occur. In 1968 less than 1 x 10^6 tonnes were measured whereas, in the following year almost 5 x 10^6 tonnes were measured. The lower line, for the period 1971-1974, shows the measured load at Red Deer. The amplitude of variation is reduced as is the total tonnage when compared to the Bindloss values. The shaded area in this instance represents incoming sediment derived downstream of Red Deer. Unfortunately, no data are available for the sediment derived between Red Deer and Drumheller to provide finer detail, but the picture is fairly self-evident. From a portion of the basin that produces only about 15 percent of the total discharge comes about 90 percent of the total suspended load.

In order, however, that some more accurate method of pinpointing the sediment supply area may be obtained, it is necessary to look at the methods by which sediment yields may be measured on the ground with particular emphasis to the situation in the Red Deer Badlands.

CRITERIA FOR IDENTIFICATION OF SEDIMENT YIELDS

In any study attempting to determine sediment yields within drainage basins a number of problems need to be overcome. These differ in degree of difficulty and

172

significance, often depending on the particular characteristics of the basin being studied (Imeson, 1974). There
are, however, four items which are more or less required
for any successful investigation. The ease, or otherwise,
with which these are solved will have a considerable bearing
on the validity of the study and the accuracy of the results.

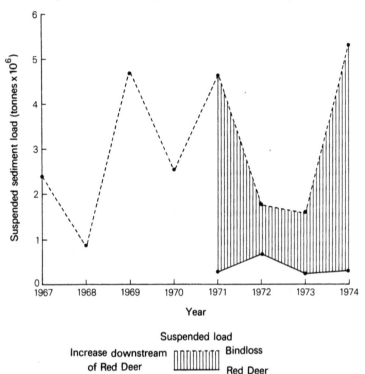

Figure 4. Annual suspended loads for Red Deer and Bindloss.

    1. The key factor is clearly the identification of
those areas which are serving as major sediment sources
within the basin.

    2. It is necessary to devise some technique(s) for
monitoring the rate and character of surface changes or
measuring the amount of sediment produced per unit area
of ground surface.

    3. Measurement of erosion rates is not particularly
useful alone without some idea of the 'sediment delivery
ratio." That is, the proportion of eroded material which

is delivered to the channel system within some specified time span.

4. Some external measurement is necessary on which to judge the accuracy of factors 1, 2 and 3 above. A sediment sampling point downstream of the assumed sediment source can provide a valuable check on the estimated rates of sediment yield. If this can be compared with sediment data collected upstream of the sediment source area, then an even more meaningful set of relationships may be determined.

In the case of the Red Deer Basin study, these problems have been solved with a high degree of satisfaction. The major sediment source area in the form of the Red Deer Badlands is clearly identified both in terms of its geographical location within the basin and its geomorphic role as a known environment of rapid erosion. Though no previous data on erosion rates existed for these particular badlands, studies of analogous regions elsewhere in western North American (Schumm, 1956; Hadley and Schumm, 1961) indicated that the Red Deer Badlands should act as a major source of sediment to the river.

Several schemes have been devised to measure both erosion rates in terms of surface change and to collect the sediment eroded from the surface. The first generally takes the form of pins, nails and washers, etc., that have been designed for a number of studies and that have been in use for many years (see for example Bridges and Harding, 1971; Emmett, 1965). The second set of devices usually involves the installation of sediment collecting troughs (Gerlach troughs, etc.) into which sediment accumulates from the area above the trough (see for example Campbell, 1970a; Soons, 1971).

The implantation of pins, and similar devices, though rapid and simple, presents some difficulties. For one thing there may be some considerable disturbance of the surface into which the pin is driven. This may enhance localized erosion about the pin giving misleadingly high rates of measured erosion. Surface flow of water and sediment about the pin may create localized scour and deposition, again distorting the

174

results. The additional factor involves taking the measurement from the head of the pin, or some mark on it, to what may be an irregular and often fragile, crusted surface. Accurate measurements under such conditions may be difficult to make.

Troughs also present problems. The installation itself is a time-consuming and often frustrating experience involving attempts at very careful sealing of the lip between the trough and the ground. The problem of maintaining adequate seals is often insurmountable on soils subject to differential expansion or contraction, either because of frost or dessication and expansion drying and wetting. Troughs also are frequently isolated as parts of closed-systems in which the trough forms part of a runoff plot which precludes sediment input from outside the system. Such closed systems may give distorted results. If closed systems are not used, one then faces the problems of delimitation of the sediment contributing area above the trough. Troughs were used in a summer only study in the Red Deer Badlands in 1968 but it soon became evident that they were at best, a temporary and not particularly satisfactory solution (Campbell, 1970a).

In recognition of these previous efforts and their various drawbacks, an erosion measuring frame ('bedstead') was designed in 1968 (Campbell, 1970b) and this has been in use ever since. It has proved highly satisfactory, gives a dense network (25 points) of highly accurate measurements (± 0.5 mm) in a one square meter area and has all the advantages of an open system technique (Campbell, 1974).

The problem of 'sediment delivery ratio' is a very real one and it is known to be a highly variable factor (Roehl, 1962). The basic problem concerns the ultimate depositional site of the eroded sediment. There is evidence from a number of studies that much of the sediment eroded from the land surface is simply redeposited on lower slopes within the basin and perhaps either never reaches the stream system, or is retained on the surface

for considerable time periods in the form of slope colluvium, fans, or a variety of floodplain deposits (Costa, 1975).

In the case of the Red Deer River this has not appeared to be a particular problem, except in one special case discussed below. Because the Red Deer Badlands are steeply sloping linear features and are very close to the river's main channel, there is very little topographic opportunity for accumulation of eroded material. Measurements do show (Table 2) that some accumulation occurs in specific localized sites, but there is no evidence to indicate that this is a general tendency. The channels formed in the badlands are steep, often deeply entrenched and exhibit no long-lived accumulation patterns of sediments (Faulkner, 1970).

What does seem to be occurring is the tendency of the Red Deer channel itself to act as a sediment storage system. The badlands, because of their steep topography, low infiltration capacity, and the flashy nature of runoff events, represent a high energy environment. Relatively small amounts of sediment-laden runoff rapidly drain the slopes. The short steep streams enter the Red Deer and discharge their load in the form of small bars and deltas. These frequently become isolated channel bars some short distance below the point of initial deposition.

Examination of the detailed long profile of the Red Deer (Figure 5) reveals a discernible 'hump' in that section of the riverbed associated with the stretch of badlands. It seems probable that normally the rate of input of material from the badlands exceeds the short-term (annual?) transport capacity of the Red Deer. Presumably occasional high floods will flush this material out since it is evident that individual high discharges do transport a considerable amount of material (Table 3).

Plot	Initial Slope Angle[1]	Surface Character	Mean Surface Change (mm)[2]
1	21° 50'	Disaggregated shale fragments 5 - 10 cm deep resting on a dense, hard, shale sub-layer of compacted shards	-23.5
2	4° 50'	Slope foot area of alluvial deposits, thin laminae of fine sand and silt	29.3
3	5° 50'	Similar to Plot 2 but with a considerably smaller amount of alluvial deposition	5.1
4	33° 00'	Hard sandy shale with three well-developed rills crossing the plot	-37.8
5	27° 00'	Dense shale, well indurated and littered with a thin layer of small (2-5 cm) clay ironstone fragments	-34.3
6	40° 20'	Highly desiccated sandy shale with a sparse covering of clay ironstone fragments	-6.9
7	47° 00'	Friable sandy shale with occasional clay ironstone fragments	-28.3
8	7° 20'	Disaggregated shale, highly "popcorned" surface over a hard clay-rich shale base	-27.4
9	13° 00'	Similar to Plot 8 but 5 m downslope of it	-63.1
		Mean of Negative Values Only	-31.6

[1] Measured on July 18, 1969; [2] Mean surface change (mm) from July, 1969 – October, 1976.

Table 2. *Surface Characteristics and Surface Changes*

177

Station	Year	Date	Daily Load (Tonnes)	Concentration (mg/1)	Daily Discharge (m³/s)
Red Deer	1971	April 15	49,950	1,460	328.6
	1972	June 26	185,810	2,750	648.7
	1973	May 28	38,312	1,390	264.5
	1974	June 18	34,538	1,460	226.9
Bindloss	1968	July 26	85,035	6,780	120.3
	1969	April 9	454,688	5,930	736.5
	1970	June 18	416,433	11,200	356.9
	1971	April 16	623,010	7,710	764.8
	1972	June 29	234,995	4,390	512.7
	1973	June 19	147,555	8,470	167.4
	1974	April 19	415,340	7,740	515.5

Table 3. Peak Daily Suspended Sediment Loads and Associated Concentration and Discharge Values (Water Survey of Canada).

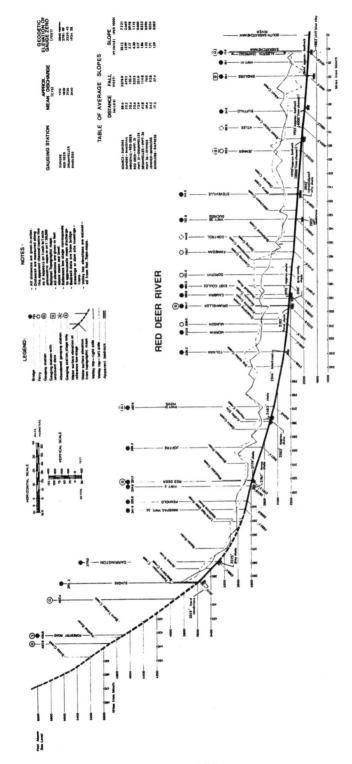

Figure 5. Long profile and topographic characteristics of the Red Deer River (Source: Alberta Research Council, 1972).

179

RED DEER BADLAND EROSION RATES

A study operating since 1969 has shown a mean annual rate of erosion of about 4.50 mm on a wide variety of surfaces and lithologies in the Red Deer Badlands (Table 2). The areal values for this erosion rate are 8,217 tonnes/km^2/yr. What this means in terms of potential sediment yield is calculated by multiplying the mean annual erosion rate value by the estimated total badlands surface area, times the specific density of the parent material:

$$4.50 \text{ mm} \times 800 \text{ km}^2 \times 1,826 \text{ kg/m}^3 = 6.58 \times 10^6 \text{ tonnes.}$$

On the average then, since 1969, each year a potential sediment supply of 6.58×10^6 tonnes is available for the Red Deer to transport. This assumes a 100 percent delivery ratio. The mean annual load for the Red Deer is a little less than 3×10^6 tonnes at Bindloss, and is about 3×10^5 tonnes at Red Deer (Figure 4). The long term average value at Red Deer will probably be less than this since 1972 was clearly an aberrant year distorting the mean upwards.

As a first approximation the figures of potential supply versus measured sediment in the river are fairly close. The potential sediment load is about twice the amount measured in the river. If allowance is made for some rather ill-defined factors then the 'error' could be considerably reduced. For instance, no adjustment is made for bedload in the Bindloss data. This may be quite high; perhaps an additional 20 percent? The suspended sediment sampling program at Bindloss is based on usually one measurement per day during the summer months and often in the winter months only one or two samples per month are taken (Kellerhals and others, 1974). It is likely that the present non-depth integrating sampling program by the Water Survey of Canada considerably underestimates the actual suspended sediment load of the Red Deer River.

The specific density values may be too high. They were based on an average bulk density sample of a representative range of unweathered lithological types present in the badlands (Campbell, 1970a). If, however, the sediments are being largely derived from a partially-weathered surficial debris, which is almost certainly true, then a specific density factor of nearer 1,000 kg/m^3 may be more appropriate. This in itself would produce almost a halving of the mass of the calculated potential supply of sediment. Finally, it is probable that the estimated area of the badlands themselves (800 km^2) may be too great, though the figure is probably as close as can be obtained without very laborious measurements from large scale air photographs.

All in all the results are good, bearing in mind the influence of all the factors involved, and they show that in this particular study a reasonable amount of accord can be obtained between ground surface measurements and suspended load sampled data. The implications are important for several reasons.

IMPLICATIONS AND CONCLUSIONS

On a very large scale the Red Deer Basin exhibits qualities analogous to those that have been identified in the 'partial area concept' (Betson, 1964; Betson and Marius, 1969). That is to say, the majority of flow in the river is derived from one relatively fixed region, the headwaters, but with variable and often highly restricted amounts from local areas, the badlands, developed largely as a result of individual storm events. In like manner, sediment origin conforms to a similar pattern. While most of the sediment in the river is derived from one general region, the badlands, because that region is relatively large and is not normally entirely covered by one storm event, each individual contribution of sediment to the river occurs as a highly variable process. Throughout the summer season, however, most if not all of the badlands will be covered on a number of occasions by a succession of storm events at different times and give the same result as a more general precipitation - runoff situation.

The fact that sediment is so overwhelmingly derived from such a comparatively minor portion of the basin, about 2 percent by area, raises additional points for consideration. As was previously mentioned, this situation is certainly not unique. Probably in most basins sediment yield comes from quite a restricted area (Gregory and Walling, 1973). Rhoades, Welch and Coleman (1975) found that 51 percent of the sediment yield in the basin they studied came from 1 percent of the area and Bowie, Bolton and Spraberry (1975) found that about 40 percent of the sediment in some northern Mississippi drainage systems came solely from channel erosion. In areas like badlands, where material is available in an unlimited supply, the sole constraint on sediment yield is the erosion mechanism (precipitation of sufficient amount and intensity to cause runoff). It is 'transport limited' (Thornes, 1976). Under these circumstances it is not surprising to find comparatively restricted areas yielding such massive amounts of sediment.

The results have considerable bearing on the concept of regional erosion rates. According to Slaymaker and McPherson (1973) the mean annual value for denudation in the Red Deer Basin lies between 0.021 and 0.04 mm/yr (clastic load only). Fournier (1960) shows the same region to have regional erosion rates of 0.0205 to 0.204 mm/yr. Both sets of data are comparable, and they should be since they are derived by the same technique, i.e., taking the suspended sediment load of the Red Deer at Bindloss and averaging it over the entire 43,000 km^2 area.

If the suspended load as measured at the Red Deer City gauge is averaged back over the watershed from which it is derived, making the simplistic assumption that it is uniformly produced, then the sediment yield for the foothills and mountains section becomes about 35 tonnes/ km^2/yr. Luk (1975) in a very detailed study found the

182

region to produce between 16 - 146 tonnes/km^2/yr. and
Strakhov (1967) in his comprehensive examination of regional values found this area of the Rocky Mountains to
have sediment yields of 50 - 100 tonnes/km^2/yr. The
three sets of values are similar and, since they are
based on different techniques, may be a reasonable range
of values for this region.

The area of basin between Red Deer and Bindloss
however, is a different proposition. Its area is about
31,600 km^2 and if the suspended sediment load at Bindloss,
exclusive of that amount which is in the river at Red
Deer, is averaged over that area, then the sediment yield
becomes about 70 tonnes/km^2/yr. The rates of erosion in
badlands show values of 8,217 tonnes/km^2/yr. and even
if they are only approximately correct, there is ample
supply of material from 2 percent of the basin to account
for the entire suspended load as measured at Bindloss.
Vast areas of basin between Red Deer and Bindloss contribute neither runoff nor sediment to the river. To refer
to regional erosion rates or sediment yields in the light
of these findings is an exercise in simplistic speculation.

REFERENCES CITED

Alberta Research Council, 1972, Hydraulic and geomorphic characteristics of rivers in Alberta: River Engineering and Surface Hydrology, Report 72-1, 52 pp. ´

Betson, R. P., 1964, What is watershed runoff?: Jour. Geophys. Res., v. 69, p. 1541-1551.

_____, and Marius, J. B., 1969, Source areas of storm runoff: Water Resources Res., v. 5, p. 574-582.

Bowie, A. J., Bolton, G. C., and Spraberry, J. A., 1975, Sediment yield related to characteristics of two adjacent watersheds: U.S. Dept. of Agric., Agric. Res. Service, ARS-S-40, p. 89-99.

Bridges, E. M., and Harding, D. M., 1971, Micro-erosion processes and factors affecting slope development in the Lower Swansea Valley: Trans. Inst. Brit. Geogrs., Spec. Pub. No. 3, p. 65-79.

Campbell, I. A., 1970a, Erosion rates in the Steveville Badlands: Canad. Geogr., v. 14, p. 202-216.

_____, 1970b, Micro-relief measurements on unvegetated shale slopes: Prof. Geogr., v. 22, p. 215-220.

_____, 1974, Measurements of erosion on badlands surfaces: Zeitschr. fur Geomorph., Supplementband 21, p. 123-137.

Costa, J. E., 1975, Effects of Agriculture on Erosion and Sedimentation in the Piedmont Province, Maryland: Geol. Soc. America Bull., v. 86, p. 1281-1286.

Emmett, W. W., 1965, The Vigil Network: methods of measurement and a sampling of data collected: Int. Assoc. Sci. Hydrol., Pub. 66, p. 89-106.

Faulkner, H., 1970, Morphology of the Steveville badlands, Alberta: Unpbl. M.Sc., thesis, University of Alberta.

Fournier, F., 1960, Climat et Erosion; la Relation entre l'Erosion du sol par l'Eau et les Precipitations Atmospheriques: P.U.F. Paris, 201 pp.

Green, R., 1970, Geological Map of Alberta: ´Map 35, Research Council of Alberta.

Gregory, K. J., and Walling D. E., 1973, Drainage Basin Form and Process: London, Edward Arnold, 456 p.

Hadley, R. F., and Schumm, S. A., 1961, Hydrology of the Upper Cheyenne River Basin: U. S. Geol. Survey Water Supply Paper 1531-B, p. 137-198.

Supply Paper 1531-B, p. 137-193.

Imeson, A. C., 1974, The origin of sediment in a moorland
     catchment with particular reference to the role of
     vegetation, in, Gregory, K. J., and Walling, D. E.,
     (eds.), Fluvial Processes in Instrumented Watersheds:
     Inst. Brit. Geogrs. Spec. Pub. No. 4, London, p. 59-72.

Kellerhals, R., Abrahams, A. D., and von Gaza, H., 1974,
     Possibilities for using suspended sediment rating
     curves in the Canadian sediment survey program:
     Studies in River Engineering, Hydraulics and Hydrology,
     Alberta Cooperative Research Program in Highway and
     River Engineering, Edmonton, Alberta, 86 p.

Luk, S. H., 1975, Soil erodibility and erosion in part of
     the Bow River basin, Alberta: Unpbl. Ph.D., thesis,
     University of Alberta.

Rhoades, E. D., Welch, N. H., and Coleman, G. A., 1975,
     Sediment yield characteristics from unit source
     watersheds: U.S. Dept. Agric., Agric. Res. Service,
     ARS-S-40, p. 125-129.

Roehl, J. W., 1962, Sediment source areas, delivery ratios
     and influencing morphological factors: Int. Assoc.
     Sci. Hydrol., Pub. 59, p. 202-213.

Schumm, S. A., 1956, The role of creep and rainwash on the
     retreat of badlands slopes: Am. Jour. Science, v. 254,
     p. 693-706.

Slaymaker, H. O., and McPherson, H. J., 1973, Effects of
     land use on sediment production, in, Fluvial Processes
     and Sedimentation: Proceedings of Hydrology Symposium
     No. 9, Nat. Res. Council of Canada, p. 158-183.

Soons, J. M., 1971, Factors involved in soil erosion in
     the Southern Alps, New Zealand: Zeitschr. fur
     Geomorph. N.F., v. 15, p. 460-470.

Stelck, C. R., 1967, The record of the rocks, in, Hardy,
     W. G., (ed.), Alberta - A Natural History: Evergreen,
     Vancouver, p. 21-51.

Strakhov, N. M., 1967, Principles of Lithogenesis, vol. 1,:
     Oliver and Boyd, London, 245 p.

Thornes, J., 1976, Semi-arid Erosional Systems: Geog.
     Papers No. 7, London School of Economics and Political
     Science, 79 p.

A COMPARISON OF OBSERVED SEDIMENT-TRANSPORT RATES WITH RATES COMPUTED USING EXISTING FORMULAS

William W. Emmett

U. S. Geological Survey
Denver, Colorado

Luna B. Leopold

Department of Geology & Geophysics
University of California (Berkeley)

ABSTRACT

Total sediment discharge of the East Fork River, Wyoming has been determined by separate measures of bedload discharge and suspended-load discharge. Data of this determination are unique because (1) bed material of the East Fork River is poorly sorted and covers a wide range of particle sizes from fine sands to coarse gravels, (2) bedload is a significant portion, about two-thirds, of total load, and (3) particle sizes of the bedload are significantly coarser than particle sizes of the suspended load. These are data from which existing sediment-transport formulas were not developed, and the applicability of existing equations to transport of coarse, heterogeneous particles has not been tested.

Measured sediment discharges of the East Fork River were compared to computed sediment discharges using five commonly-accepted, empirically-derived equations: Maddock (1976), Ackers and White (1973), Yang (1973), Shen (1972), and Engelund (1967). The comparison shows an inapplicability of existing transport equations, at least those tested, to correctly predict sediment-transport rates as particle sizes and degree of heterogeneity of particle sizes increase from the values used in equation derivation.

The Ackers-White formula was selected to illustrate what particle-size parameter is required to have the equation correctly predict or agree with observed sediment discharges. For the East Fork River, using the size distribution of bed material as a basis for particle availability, the required particle size can be roughly approximated by d_{65} for flows with a sediment concentration of up to about 100 mg/L (also coincident with bankfull discharge), by d_{35} for flows with a sediment concentration greater than about 500 mg/L (flows greater than about twice bankfull discharge) and by a smooth, S-shaped transition of particle sizes for flows intermediate of bankfull to twice bankfull discharge.

GEOMORPHIC RESPONSE OF CENTRAL
TEXAS STREAM CHANNELS TO
CATASTROPHIC RAINFALL
AND RUNOFF

Peter C. Patton

Department of Earth and
Environmental Sciences
Wesleyan University

Victor R. Baker

Department of Geological Sciences
University of Texas at Austin

ABSTRACT

In central Texas the morphology of the streams incised into the Cretaceous limestone bedrock of the Edwards Plateau is controlled by catastrophic floods. This is partly a response to a climatic regime which has produced near record intensity rainfalls of up to 24 hours duration. Upstream from the Balcones Escarpment the high relief, thin lithosols, and sparse vegetation enhance the hydrologic response of the drainage basins. Flooding on two small streams in 1972, resulting from 406 mm of rainfall in four hours, produced spectacular erosion and transport of limestone bedrock. Formation of scour holes and deposition of large gravel bars significantly altered the morphology of the streams. Radiocarbon dates of buried floodplain sediments indicate a minimum recurrence interval of 400 years for geomorphically significant flooding on these streams.

Based on the lack of erosion during floods at the channelfull stage it is proposed that the formation of valley meanders is controlled by floods which greatly exceed the channelfull stage. This is substantiated by the erosion during floods which fill the entire valley and by the statistical correlation of valley meander wavelength with the maximum discharge of record.

Floodplain formation does not prohibit the formation of valley meanders. Rather, on streams incised into bedrock discontinuous floodplains really are gravel bars deposited and eroded during extremely large floods.

Streams flowing on granitic rocks in the Llano region and on coastal plain sediments downstream from the Balcones Escarpment have a morphology adjusted to frequent low-magnitude discharges. This is a function of smaller source rock sediment size, lower relief, and the increased vegetation on more permeable soils causing a reduction in the hydrologic response of the drainage basins. This indicates that, although climate may provide the potential for catastrophic stream behavior, geologic controls may have an overriding influence.

INTRODUCTION

It has long been recognized that central Texas lies within one of the most severe rainfall-runoff regimes in the conterminous United States. As early as 1917 central and northeastern Texas were delineated as regions having the greatest storm magnitudes in the eastern United States (Miami Conservancy District, 1917). Data on the intensity and duration of rainfalls indicate that many Texas storms approach world record intensities for durations up to 24 hours (Figure 1). As a result, central Texas is characterized by floods that approach, or are, U. S. records for unit-area runoff magnitudes (Hoyt and Langbein, 1955). Because of these climatic conditions, central Texas is an exceptional region in which to observe the morphological response of stream channels to extreme high-magnitude runoff events.

Furthermore, there is increasing evidence that rare great floods are significant geomorphic agents in many climatic and physiographic regions in extending the drainage network (Hack and Goodlett, 1960; Williams and Guy, 1973) forming stream terraces and floodplains (Hack and Goodlett, 1960; Gupta, 1975) transporting coarse

bedload sediment (Stewart and LaMarche, 1967) and in
eroding valley meanders (Tinkler, 1971). These studies
sharply contrast with studies, primarily of humid regions,
which indicate that frequent low-magnitude runoff events
are more important in forming and maintaining stream morphology (Wolman and Leopold, 1957; Dury, 1964). In fact,
large floods in low-relief humid regions rarely cause more
than slight amounts of channel scouring and widening, the
effects of which are usually reworked in a relatively short
period of time (Jahns, 1947; Costa, 1974; Moss and Kochel,
in press).

The study of channel morphology of central Texas
streams in relation to catastrophic floods will help
to clarify the role that high-magnitude infrequent runoff
events have in controlling fluvial morphology. These
observations will aid in evaluating the work accomplished
by processes of differing magnitude and frequency. In
addition, analysis of rare great floods and their effect
on the landscape in central Texas is important to the
understanding of the evolution of fluvial landforms in
other regions where catastrophic floods may be historical-
ly rare but geologically important.

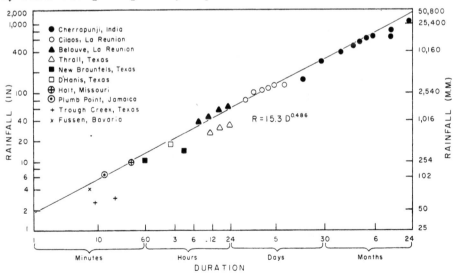

Figure 1. *Graph of world record rainfall intensities for specific durations. Many central Texas rainfalls approach world record intensities for durations of up to 24 hours.*

191

PHYSIOGRAPHIC AND METEOROLOGIC
CONTROLS ON FLOODING

The large flood magnitudes in central Texas are dependent on the physiography and climate of the region. The physiography can be separated into two regions: the Edwards Plateau and the Gulf Coastal Plain (Figure 2).

The Edwards Plateau, a limestone upland, is bounded on its southern and southeastern margins by the Balcones Escarpment, a zone of monoclinal folding and faulting which forms the interior edge of the coastal plain (Sellards, 1934). The escarpment is most pronounced between Uvalde and Austin, with the greatest relief, 300 meters, near Del Rio, Texas. Upstream from the escarpment the streams are incised into the Upper Cretaceous Edwards and Glen Rose Limestones. The high relief, low infiltration capacity of stoney soils on hillslopes and interfluves, high drainage density, and sparse vegetation contribute to the rapid runoff and high flood peaks.

Downstream from the escarpment the major streams flow in large alluvial valleys incised into the relatively unconsolidated Tertiary and Quaternary fluvial-deltaic sediments of the Gulf Coastal Plain. The lower relief, coarser drainage texture and denser vegetation dampen the hydrologic response of the streams to catastrophic rainfall. This results in more attenuated runoff hydrographs compared to the hydrographs of streams incised into the Edwards Plateau (Baker, in press).

In central Texas the climate ranges from semi-arid along the Pecos River valley to subhumid along the Colorado River near Austin. The climate is variable and rainfall is erratically distributed in time and space. The annual precipitation has varied by more than 25 percent from the average 14 times between 1931 and 1960 (Carr, 1967). World record rainfalls for durations up to 24 hours have occurred in the region (Figure 1) and it is not uncommon for the mean annual

192

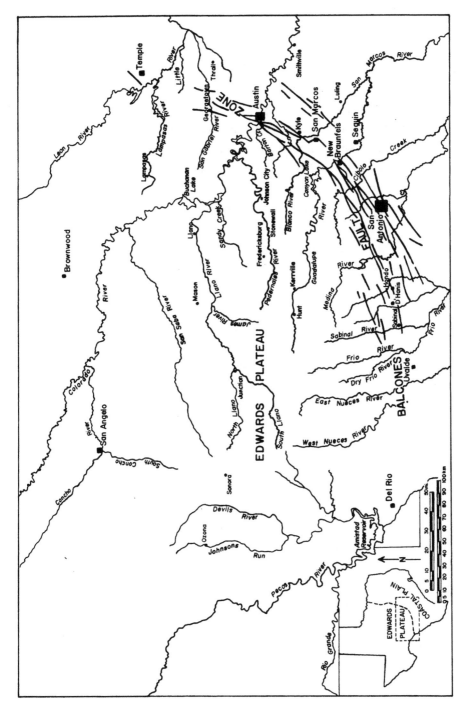

Figure 2. Location map of central Texas showing major streams and principal towns.

rainfall to occur during a single storm. A variety of
meteorological conditions are responsible for generating
the storms that produce record Texas floods. These include, a) thunderstorms, b) Pacific frontal systems,
c) broad frontal low pressure systems, known as easterly
waves, and d) hurricanes.

Thunderstorms formed by convection can reach catastrophic size in central Texas resulting in intense rainfall and hail. One thunderstorm at D'Hanis, Texas,
produced rainfall of 560 mm in two hours and 45 minutes,
a world record intensity for that duration.

Hurricanes which move inland forming extratropical
storms can cause severe flooding. Rainfall from hurricane
Alice in 1954 over the Pecos River and Devils River
basins resulted in the greatest historical flood peak
in Texas.

The predominant cause of Texas floods is low pressure systems which flow out of the easterly trade winds.
These air masses are warmed and saturated with moisture
after moving over the Gulf of Mexico. When the waves
move inland and are orographically lifted by the Balcones
Escarpment, intense precipitation results. The waves
can be additionally lifted by riding over northern cold
fronts (Orton, 1966). The close spacing of mean annual
isohyets along the escarpment (Carr, 1967) and the
strong parallelism of actual storm isohyets to the escarpment (Dalrymple, 1937) indicate the escarpment's influence
in controlling rainfall distribution. The famous Thrall,
Texas storm of September 9-10, 1921 in which 825 mm of
rain fell in 18 hours and the disastrous September 1952
floods on the upper Colorado River (660 mm in 54 hours)
were caused by the incursion of easterly waves over
central Texas.

Central Texas rainfall regime produces an erratic response in the streams. Periods of low runoff are interrupted by enormous floods. A classic example of this phenomenon is the Alice Hurricane and resulting flood on the

Pecos River. The hurricane centered over the Pecos-Devils River divide and dropped up to 1,067 mm of rain during the period June 24-29 (U. S. Soil Conservation Service, 1954). The resulting flood peak on the Pecos River near Comstock, Texas was 27,440 m³/sec (980,000 cfs) nearly eight times the previous recorded maximum discharge. At Comstock the water stage was 29 meters, more than 17.5 meters above the previously recorded maximum flood stage. It is even more amazing that just 24 kilometers upstream at Sheffield, Texas the maximum discharge for the same period was only 475 m³/sec. Therefore, the contributing area for the flood was only 9,300 km².

It is not possible to accurately estimate the frequency of this flood based solely on the available hydrologic data. There is, however, a Holocene flood history of the Pecos River preserved in an early man site located near the confluence with the Rio Grande. Arenosa shelter is a bedrock overhang, probably formed by the Pecos as it incised into the limestone walls of the valley. About 9,500 years ago (University of Texas Radiocarbon Laboratory, Date Tx-668) the river was diverted and overbank sediments began to accumulate in the overhang. Within the shelter, alluvial sediments are interstratified with layers containing ash, charcoal, and artifacts. Analysis of the alluvial stratigraphy combined with radiocarbon dates of the cultural debris provide a means of establishing a paleoflood frequency record for the Pecos River (Patton, 1977). The Alice Hurricane flood buried a surface that was 1,300 years old and the archeological evidence in the shelter indicates that the flood has a recurrence interval in excess of 2000 years (Figure 3).

STREAM CHANNEL MORPHOLOGY

The bedrock controlled streams in central Texas differ in morphology as a result of bedrock type and source rock sediment size and availability. Shepherd (1975) classified bedrock streams in central Texas

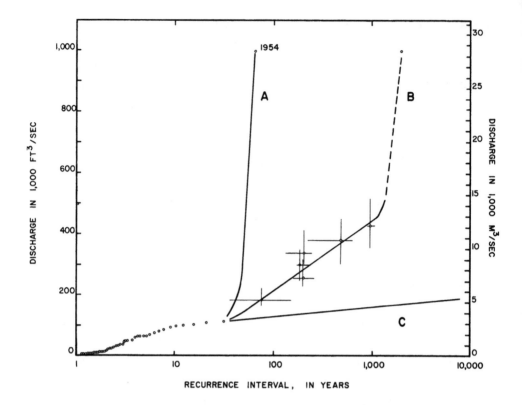

Figure 3. *Flood frequency curves for Pecos River near Comstock, Texas.
Curve A, flood frequency curve including the 1954 flood with
the other 53 years of record. Based on this record the 1954
flood would have a recurrence interval of 55 years. Curve C
is the extrapolated pre-1954 flood frequency curve. Based
on this curve the 1954 flood would have a recurrence interval
of millions of years. Curve B is the paleoflood frequency
curve estimated from the alluvial stratigraphy. Vertical
bars represent possible errors in estimating discharge relative to stages of deposition, and horizontal bars indicate
the standard deviations of the radiocarbon dates. The alluvial chronology refines the flood frequency between 100 and
1,000 years.*

based on these criteria in a manner similar to the classification of alluvium streams by Schumm (1963). One end member of the spectrum is represented by streams draining granitic terrain in the Llano region; these streams are wide, shallow, relatively straight and have an abundant bedload of well sorted sand which commonly fills the entire valley up to the bedrock walls. The abundance of sand-size alluvium is a result of the grusification of the Llano granites in the present climatic regime. Because the bedload is entrained during all runoff events (Shepherd, 1975), infrequent large floods do not exert a significant influence on the channel morphology. Erosional features created in the bed of the stream during high flows are rapidly obliterated during low flow conditions.

The other end of the spectrum of bedrock channel types is represented by streams incised into limestone. These streams are characterized by narrow, deep, channels which have a well-developed pool-riffle pattern that exposes the bedrock floor of the channel (Shepherd, 1975). The channels are sinuous and the meanders are incised into the Edwards and Glen Rose Limestones, forming broad slipoff slopes on the inside of meanders and steep scarps on the outside of the bends. That the streams are actively eroding these meanders is evidenced by the undercutting of the steep scarps contributing very coarse sediment to the active channels.

Within the channels, bars composed of coarse gravel occur as point bars and alternate bars. Bars also occur in channel expansions and downstream from bedrock projections into the channel (Baker, in press). On Cibolo Creek near Bulverde, Texas, long reaches of channel eroded to bedrock are separated by shorter reaches choked by coarse limestone alluvium. The gravel in these bars is probably only entrained by infrequent large floods. With the exception of transporting fine-grain sediment, which eventually masks coarse gravel bar forms and allows vegetation to become established, more frequent low flows have little effect on

197

the morphology of the streams.

EROSION AND SEDIMENTATION
ON LIMESTONE STREAMS

Observations of erosion and deposition following a
flood near New Braunfels, Texas provide insight into the
processes governing the morphology of small high-gradient
streams incised in limestone. The flood resulted from a
line of severe thunderstorms on May 11 and 12, 1972 which
migrated northeastward along the Balcones Escarpment
and caused an average rainfall of 200 mm over an area
of 800 km^2. The Soil Conservation Service (Colwick and
others, 1973) reported a maximum rainfall intensity of
305 mm/hr and a maximum rainfall of 406 mm for the storm.
The heaviest rainfall was centered over the Guadalupe
River and two small tributaries, Elm Creek and Blieders
Creek. The peak discharge on the Guadalupe River at
New Braunfels was 2,830 m^3/sec (100,000 cfs), equalling
the historical maximum discharge at this station. The
total runoff was generated from a contributing area of
about one-fifth the total area of the Guadalupe basin and
had a recurrence interval of about 35 years.

On smaller streams the runoff was staggering. An
indirect discharge measurement on Blieders Creek estimated peak discharge at 1,360 m^3/sec (48,000 cfs) from
a 39 km^2 drainage area; about 36 m^3/sec/km^2 (3,200 cfs/
mi^2). Peak runoff in Elm Creek was estimated at 1,130 m^3/
sec (40,000 cfs) or nearly 87 m^3/sec/km^2 (8,000 cfs/mi^2)
a record Texas unit-area discharge magnitude.

There was little or no erosion along the Guadalupe
River during this flood; however, along the smaller
tributaries the erosion and deposition was spectacular.
The greatest effects of flooding were observed in the
lower reaches of Elm Creek, an ephemeral stream with a
drainage area of only 12.5 km^2 in the dense well jointed
Edwards Limestone. Prior to the 1972 storm and flood,
Elm Creek's channelway was almost completely obscured

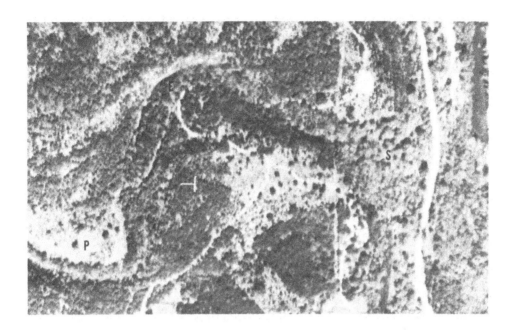

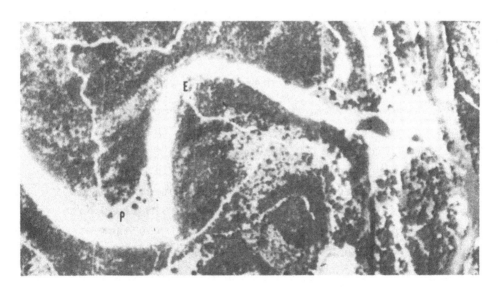

Figure 4. Aerial photos of Elm Creek channel before (above) and immediately after (below) the 1972 flood. The two photos have slightly differing scales, but both cover areas of 400 x 800 m. Channel scour, vegetation uprooting, and gravel deposition can be seen by comparing common points on the two photos. P is point bar, E is exposure shown in Figure 5, S is a scour hole.

199

by vegetation (Figure 4, upper photograph). After the
flood, the lower reaches were marked by destruction of
vegetation, intense scour, and deposition of gravel ranging
from pebbles to huge boulders (Figure 4, lower photograph).

Bedrock erosion was localized at discrete scour holes,
with the greatest scour at a meander. At that point a 30
by 100 meter pool was cut approximately three meters deeper
than the elevation of the upstream and downstream riffles
(Baker, in press). In the riffles between the scour holes
boulders up to 2.5 x 1.5 x 1 meter were deposited. These
large boulder were derived from the erosion of the scour
holes and from blocks that had spalled off the steep
limestone valley walls during periods between floods.
Much of the gravel transported by Elm Creek during this
flood was deposited as a large bar at the mouth of the
creek where it flows into the Guadalupe River.

On Blieders Creek an alternating sequence of scour
holes and gravel-choked riffles were formed and a majority
of the trees in the valley were either uprooted or broken
off at the base. As in Elm Creek, the most intense scour
was located at meander bends where the flow was constricted.

FREQUENCY OF CATASTROPHIC FLOODS

An estimate of the frequency of channel-modifying
floods on central Texas streams, such as the 1972 flood
on Blieders Creek and Elm Creek, would be a valuable
comparison to channel forming discharges in other climatic-
physiographic regions. However, it is not possible to
precisely estimate the frequency of the 1972 flood with
standard hydrologic techniques because of the lack of
long-term records in the region. Based on rainfall
frequency studies, the storm that caused the New Braunfels
flood was about twice the intensity of the 100-year return
period rainfall (Colwick and others, 1973) but trans-
lating rainfall frequencies into runoff frequencies is
inexact.

Fortunately, there is geomorphic evidence for the frequency of flooding on Elm Creek and Blieders Creek. Erosion of
their floodplains exposed stratigraphic sequences of flood-
deposited gravels separated by layers of overbank sediments on
which organic rich soils had developed. Radiocarbon dates of
the buried soils provide an approximate estimate of the recurrence interval for major floods. On Blieders Creek the buried
soil is exposed in a chute eroded through a point bar during
the 1972 flood. The buried soil is thin and irregular in
thickness and may have been partially eroded during the flood
that buried it with a thin veneer of gravel. A modern soil
has formed on 60 cm of overbank sediment deposited since this
flood. The 1972 flood buried the entire sequence with another
layer of gravel. Therefore, there are three floods represented
in the section, separated by two periods of soil formation.

On Elm Creek a similar stratigraphy is visible in an eroded floodplain deposit. A thick buried soil is bracketed with
large boulders and the entire sequence is mantled with a layer of gravel deposited in 1972 (Figure 5). The buried soils
on both streams and the base of the surface soil on Blieders
Creek were sampled for radiocarbon dating. The age of the
buried soils on both Elm Creek and Blieders Creek will give a
minimum estimate for the total time period in which the two
most recent floods occurred. An age on the base of the surface soil on Blieders Creek will give a minimum time interval
between these same two floods.

There are significant problems associated with radiocarbon dating of soils. The radiocarbon dating technique gives
the elapsed time since the death of the organisms, but the
accumulation of dead organic matter in soils is a continuous
process. A date obtained from soil will be the result of the
entire period of accumulation and will not date the initial
time of soil formation (Schapenseel, 1971). The oldest date
obtained from a soil, therefore, represents a minimum age of
soil formation.

Soil humus including the soluble humic acid and the insoluble residue, were dated. The acid is the most likely
source of contamination in the soils, whereas, the insoluble

201

residue should most closely represent the mean residence time of the soil (Geyh and others, 1971). Uncorrected dates for the two fractions and the total soil humus are listed in Table 1. The soil residue dates for the buried soils are comparable and suggest that burial may have occurred at the same time. This is logical because the drainage basins are only several kilometers apart. Since each buried soil is capped by flood gravels and both sequences are capped by gravels deposited by the flood of May 11 and 12, 1972, the two floods represented have occurred during a time interval of about 700-900 years. An average minimum recurrence interval for the floods of about 400 years is suggested. The recurrence interval estimate is geomorphically important in that it places a minimum interval on flows which are capable of drastically eroding the bedrock channels and controlling the larger aspects of the stream channel morphology.

On limestone streams recovery time following severe erosional floods is long. Unlike alluvial streams in the eastern U. S., where the flood effects of the 1972 Agnes flood were completely modified by lower flow durations in one year (Costa, 1974) in central Texas it probably requires several years before even vegetation becomes re-established. Reworking of the gravel flood bars and infilling of the scour holes is probably not accomplished for an even greater period of time, perhaps tens of years.

Supporting evidence for this view can be seen on many central Texas streams. Cibolo Creek, a tributary of the San Antonio River, is an excellent example. In 1964 the stream experienced a severe flood, the largest flood since 1892 at Boerne, Texas (Rostvedt and others, 1970). Where erosion and transportation of sediment was greatest, the flood effects are still visible. In 1976 little vegetation had re-established in scoured areas and many of the scour holes still had their post-flood morphology preserved. Areas adjacent to the channel that were buried by gravel are not revegetated and other reaches of the channel are still

202

Location	Laboratory no.	Age in Years B.P.	Material Dated
Upper Soil Blieders Creek	Tx – 2348	710 ± 70	Total Soil Humus
		770 ± 90	Humic Acid Fraction
		370 ± 40	Insoluble Organic Residue Fraction
Buried Soil Blieders Creek	Tx – 2349	1000 ± 60	Total Soil Humus
		790 ± 150	Humic Acid Fraction
		890 ± 60	Insoluable Organic Residue Fraction
Buried Soil Elm Creek	Tx – 2350	1300 ± 60	Total Soil Humus
		950 ± 70	Humic Acid Fraction
		710 ± 50	Insoluble Organic Residue Fraction

Age calculations based on ^{14}C half life of 5,568 years and modern reference standard of 95% of National Bureau of Standards oxalic acid. Ages given to one standard deviation.

Table 1. Radiocarbon Ages of Floodplain Soils on Blieders Creek and Elm Creek, Texas.

choked with large boulders derived from the 1964 scour holes.

FORMATION OF VALLEY MEANDERS

The erosion in constrictions at meander bends that occurred during the 1972 flood is clear evidence that bedrock valley meanders are also eroded during brief infrequent periods of catastrophic runoff. Unlike alluvial rivers which have a meander wavelength correlated with their bankfull discharge (Dury, 1964) bedrock valley meander wavelength is undoubtedly related to discharges in excess of the one to two year frequency commonly associated with bankfull conditions. The formation of valley meanders and floodplains is a geomorphically important response of central Texas streams to infrequent catastrophic floods.

A previous investigation of the active valley meanders in central Texas (Tinkler, 1971) concluded that the valley meander wavelengths were related to erosion during floods in which the flow completely filled the bedrock channel (i.e., channelfull discharge). Tinkler reasoned, that for discharges up to the confines of the bedrock channel banks the depth of flow in the channel would increase rapidly with increasing discharge. At stages above the limits of the channel the rate of increase in depth would diminish. Since depth of flow can be roughly related to the shear stress on the bed of the stream, channelfull discharge would then represent a maximum effective depth and a limit on the powers of corrasion of the stream. He estimated that channelfull discharge on central Texas streams had a recurrence interval of 10 to 50 years.

Tinkler stated that the primary controls on the development of valley meanders were, (1) the heavy storm rainfall and high, rapid runoff and (2) the absence of significant floodplain development, although incipient floodplains could be found locally along streams in the

204

Edwards Plateau. He suggested that once floodplains formed, an apparent reduction in effective discharge occurred which caused the valley meanders to become fossilized. He hypothesized that the formation of floodplains along central Texas streams resulted from an increased sediment yield as a function of increased surface area exposed during the incision of the plateau.

Based on measurements of many central Texas bedrock stream channels, we suggest that the development of floodplains is more common and more complex than Tinkler suggests and that Tinker's estimated recurrence intervals for floods responsible for eroding the valley meanders of small streams may be too conservative.

First, the development of floodplains adjacent to low-water stream channels is not uncommon on streams incised into the limestone plateau. The cross-section morphology of stream channels on the plateau can be roughly divided into four types: (1) streams which have discontinuous floodplains of coarse bouldery alluvium; (2) streams which have incised bedrock inner channels, similar to those described by Shepherd and Schumm (1974), forming a relatively flat bedrock surface adjacent to the channel; (3) broad, deep, channels bordering wide slip-off slopes which limit the depth of flow in the channel; and (4) narrow canyons in which flow depths rapidly increase for all conceivable increases in discharge. The Pedernales River downstream from Johnson City is an example of this last channel type which Dury (1964) described as a natural flume.

In this discussion we are primarily interested in the first channel type. We have already noted the aggradation of coarse alluvium adjacent to the main channel during catastrophic floods in this region. This aggradation can be observed in exposures of older floodplain deposits and in gravel bars and surfaces deposited by historical floods. The accretion

of these flood deposits restricts the low-flow channel
and, therefore, channelfull discharge occurs more frequently. In fact, on the basis of available hydrologic
data, channelfull discharge on central Texas streams can
have a recurrence interval of less than ten years
(Patton, 1977). Furthermore, channelfull flows do
little or no erosion of the floodplain or the bedrock
meanders. For example, during the 1972 New Braunfels
flood the Guadalupe River filled its channel and covered
its floodplain with several feet of water, but with the
exception of a few minor scour holes, no significant
erosion was observed.

It is most important to note that these coarse
alluvial floodplains are not stable. Just as these
forms are created by catastrophic floods so they can be
eroded during floods of equal magnitude. Obviously,
the time period between deposition and removal is long,
perhaps several hundred years, as evidenced by the
accumulation of overbank sediments and the establishment of vegetation. On small streams in central Texas
it has already been estimated that periods of floodplain stability may last about 400 years.

As a final test of Tinkler's hypothesis that channelfull discharge represents the effective discharge which
erodes and forms the valley meanders, additional meander
wavelength data were collected using the method described by Dury (1958). The streams sampled had either
bedrock banks or beds and when plotted as a function
of drainage area, all but one wavelength value fell
within the range of Dury's (1965) valley meander grouping.
Using regression techniques the wavelength data were
compared with discharge values having recurrence intervals
of 1.2 years (approximately bankfull frequency) 2.33 years
(mean annual flood frequency) and the maximum discharge
of record (the closest approximation to discharges which
fill the entire valley). The recurrence interval of
the maximum discharge of record varied from 10 to 200

years, although most of these estimates are largely a function of the short hydrologic record. The example of the difficulty in determining the flood frequency of the Alice hurricane flood on the Pecos River indicates that these values of flood frequency should not be taken too seriously. The wavelength data were also compared with channelfull discharge. Channelfull discharge data were generated by channel surveys and slope-area discharge calculations and also from recalculation at the channelfull stage of slope-area measurements on file at the Austin District Office of the United States Geological Survey, Water Resources Division. Admittedly approximations, these values were considered appropriate for comparison with maximum discharge values as most of that information was collected in the same manner.

The results of the regression analysis (Table 2) clearly indicate that the historical maximum discharge best explains the variability in meander wavelength. This is not a precise relationship but it does suggest that great floods, in many cases an order of magnitude greater than channelfull discharge, are responsible for the erosion of the valley meanders. Therefore, although the formation of smaller channels by bar deposition and floodplain formation increases the frequency of channelfull discharge, it probably does not retard the erosion of the valley meanders. In fact, based on evidence of historical flood stages, it can be demonstrated that flow depths can be more than double the stage of channelfull discharge (Figure 6). It is probably these flows of low frequency that are responsible for forming the bedrock channel morphology.

The sequence of floodplain formation and destruction in central Texas is similar to the evolution of

Figure 5.   *View of a portion of the eroded floodplain on Elm Creek.   The cut exposes flood gravels separated by a thick buried soil.*

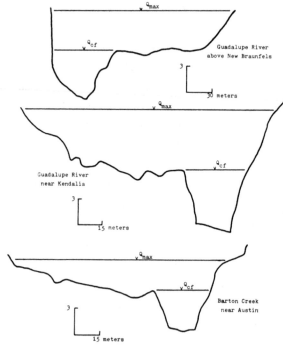

Figure 6.   *Channel cross-sections of the Guadalupe River and Barton Creek. Channelfull discharge (Qcf) on Guadalupe River above New Braunfels recurs approximately every 10 years.*

Discharge Variable	Meander Wavelength (L) Equation	Sample Size	Standard Error	r^2
Bankfull discharge $Q_{1.2}$	$L = 1996\ Q_{1.2}^{.13}$	21	.422	.06
Mean annual flood $Q_{2.33}$	$L = 30.7\ Q_{2.33}^{.61}$	21	.322	.45
Maximum recorded discharge Q_{max}	$L = 22.0\ Q_{max}^{.50}$	21	.186	.78
Channelfull discharge Q_{cf}	$L = 9.4\ Q_{cf}^{.65}$	24	.322	.58

Table 2. Least Squares Regression Equations of Valley Meander Wavelength to Discharge.

floodplains during extraordinary floods on northern California high-gradient streams (Stewart and LaMarche, 1967). During a 1964 flood on Coffee Creek in Trinity County, California, up to 1.5 meters of gravel were vertically accreted on to floodplains adjacent to the main channel. The presence of other, older flood bars within the stream valleys in northern California is evidence that this process of catastrophic erosion and deposition is not an isolated event but is the primary way in which floodplains are created in that area (Helley and LaMarche, 1973). This mechanism of floodplain formation is in direct contrast to floodplain formation in low-relief humid regions. Here studies indicate that high frequency floods create the floodplain by gradual lateral accretion of point bar deposits and that on these streams little or no erosion is accomplished during extreme floods (Wolman and Leopold, 1957). Clearly, different climatic and geologic factors are responsible for the diversity of stream response.

FACTORS CONTROLLING STREAM RESPONSE
IN CENTRAL TEXAS

The degree to which infrequent floods exert a significant control on stream channels in central Texas is a function of interrelated climatic and geologic variables. Important variables include those which in part control, (1) the temporal and spatial distribution of rainfall, rainfall intensity and duration; (2) the processes and rates of weathering which govern the type and availability of sediment supplied to the stream, and (3) soil formation and type and density of vegetation which play a major role in defining the interception, infiltration, and overland flow stages of the hydrologic cycle. The effect of these variables on stream response can be best demonstrated by comparing streams on either side of the Balcones Escarpment.

Upstream from the escarpment, stream channel response is enhanced by the high relief and limestone bedrock. Within the present climatic regime, there is only minimal chemical weathering of the limestones which retards soil formation and the establishment of vegetation. The combined effect is to have a large component of overland flow relative to interflow. Unit hydrographs for streams above the escarpment are, therefore, peaked with small lag times (Baker, in press).

In a study of regional flood potential (Patton and Baker, 1976) we have suggested that there is a feedback mechanisms which enhances drainage response in regions of intense but infrequent rainfall. The erratically distributed rainfall does not promote vegetation or soil formation and, therefore, there is a greater amount of overland flow which erodes and forms more rills and channels in the basin. This increases the drainage density which increases the efficiency of the drainage network for carrying runoff. Therefore, greater flood peaks result.

Most of the sediment supplied to the bedrock stream channels is derived from mechanical weathering and masswasting. This results in coarse debris that can only be moved during great floods as the lower flow components of the streams are unable to entrain the sediment. The large difference between the greatest flood on any stream and its mean annual flood exaggerates this effect, particularly on streams which have ephemeral flow. On larger streams with a base flow component it is quite probable that, in terms of total weight, more suspended and dissolved sediment is transported by frequent events than the total bedload sediment transported during catastrophic floods. On these streams there is a disparity between the quantity of sediment transported by a particular flow duration and the impact of that flow on the stream morphology. Here, the erosion and sedimentation during infrequent flows control the channel morphology without accomplishing the most sediment transport.

211

Downstream from the Balcones Escarpment the hydrologic and geomorphic response of the stream channels and
drainage basins is greatly different and this change
is primarily a function of geology and relief. On the
coastal plain the thicker soils developed on the clastic
sedimentary deposits combined with the lower relief and
drainage density permit interflow to become a greater
component of the total runoff. Unit hydrographs for
small streams on the coastal plain have greater lag
times and lower hydrograph peaks than streams on the
Edwards Plateau (Baker, in press).

Major streams south of the escarpment flow in
broad valleys which are filled with relict alluvium.
The bedload in these streams is easily transported by
flows in the range of bankfull discharge. As a result,
these channels show a much more sensitive response to
subtle climatic variations. In particular, paleochannel
patterns on terrace levels of the Colorado River (Texas)
show that this large river adjusted its sinuosity, wavelength, and sediment type in response to an alternating
arid-humid climate during the late Quaternary (Baker
and Penteado, 1975). In fact, the present Colorado,
reacting to a decrease in discharge in the late Holocene,
has incised into its meandering pattern and is now an
"Osage type" underfit stream.

Upstream from the escarpment there is no evidence that these climatic changes had any significant
effect on the stream channel morphology or valley
meanders. Although there is evidence of increasing
aridity during the late Quaternary causing erosion
and removal of soils from the upland surfaces on the
Edwards Plateau (Lundelius, 1967) little effect has
been preserved in the gross morphology of the streams.
It is possible that all evidence of former stream patterns
may have been removed by erosion but we speculate that

212

limestone bedrock streams on the plateau may have been
less sensitive to climate changes which affected the high
frequency end of the discharge spectrum than they were
to the continued infrequent occurrence of intense rainfall-
runoff floods.

Wolman and Miller (1960) noted that in regions of
seasonal or continuous aridity infrequent high-magnitude
processes had a relatively greater effect on fluvial morphology; particularly with regard to sediment transport.
In this manner climate does establish the potential for
catastrophic response. However, the examples from central
Texas, specifically in the Llano region and downstream
from the Balcones Escarpment, demonstrate that geology
can have an overriding control on climate in dictating
stream channel response.

SUMMARY

Stream channels incised into limestone on the Edwards
Plateau have a morphology adjusted to flows of low
frequency but extreme magnitude. Evidence of scour
and deposition on two small streams resulting from
catastrophic flooding indicate that the channel configuration and floodplain morphology are related to
these events. Radiocarbon dates indicate a minimum
recurrence interval of about 400 years for channel-
forming discharges on small central Texas streams.

The active valley meanders are also eroded during
these catastrophic floods. The lack of erosion by
floods with recurrence intervals of 10 to 50 years
suggests that only those flows which completely fill
the bedrock valleys are capable of forming the bedrock
meanders. The fact that the maximum recorded discharge
rather than channelfull discharge is statistically
better correlated with meander wavelength substantiates
this observation.

Floodplains on the smaller central Texas streams
incised into the limestone plateau are not stable

features but are created and destroyed by extraordinary floods. Although floodplain formation does restrict the low flow channels of the streams and does increase the frequency of channelfull discharge, it probably does not retard the erosion of the valley meanders.

Regions of continuous or seasonal aridity possess many of the requisites for catastrophic response. Streams in such regions tend to have coarse sediment loads and are subjected to ideal flood-frequency conditions. However, the example from Texas demonstrates that climate does not completely dictate the nature of catastrophic response. Instead, the conditions for catastrophic response are in part controlled by the hydrologic nature of the regolith, hillslope processes, bedrock characteristics, and the position of streams in the drainage network hierarchy.

ACKNOWLEDGMENTS

Studies of flood effects on Texas streams were supported by National Aeronautics and Space Administration under contract NAS 9-13312, The University of Texas Bureau of Economic Geology, The Geology Foundation of The University of Texas at Austin, and The Geological Society of America through a Penrose Research Grant, no. 1947-75. R. Foutch, S. D. Hulke, and E. Patton assisted with the field work. Elsie Patton edited and typed several drafts of this manuscript.

REFERENCES CITED

Baker, V.R., (in press), Stream channel response to floods
    with examples from central Texas: Geol. Soc. America
    Bull.

_____, and Penteado, M.M., 1975, River adjustments to late
    Quaternary hydrologic regimen changes in central Texas:
    Geol. Soc. America Abstracts with Programs, v. 7, no. 2,
    p. 144.

Carr, J.T., 1967, The climate and physiography of Texas:
    Texas Water Dev. Board Rept. 53, 27p.

Colwick, A.B., McGill, H.N., and Erichsen, F.P., 1973,
    Severe floods at New Braunfels, Texas, May 1972: Am.
    Soc. Agric. Engineers, paper presented at 1973 annual
    meeting, Lexington, Ky., 6p.

Costa, J.E., 1974, Response and recovery of a piedmont water-
    shed from tropical storm Agnes, June 1972: Water Resour-
    ces Res., v. 10, p. 106-112.

Dalrymple, T., 1937, Major Texas floods of 1936: U.S. Geol.
    Survey Water Supply Paper 816, 146p.

Dury, G.H., 1958, Tests of a general theory of misfit
    streams: Trans. Inst. Brit. Geog., Pub. 25, p. 691-706.

_____, 1964, Principles of underfit streams: U.S. Geol.
    Survey Prof. Paper 452-A, 67p.

_____, 1965, Theoretical implications of underfit streams:
    U.S. Geol. Survey Prof. Paper 452-C, 43p.

Geyh, M.A., Benzler, J.H. and Roeschmann, G., 1971, Problems
    of dating Pleistocene and Holocene soils by radio-
    metric methods: in, Yaalon, D.H., (ed.), Paleopedology,
    origin, nature and dating of paleosols: Intern. Soc.
    Soil Sci. and Israel Univ. Press, p. 63-75.

Gupta, A., 1975, Stream characteristics in eastern Jamaica,
    an environment of seasonal flow and large floods: Am.
    Jour. Science, v. 275, p. 825-847.

Hack, J.T., and Goodlett, J.P., 1960, Geomorphology and
    forest ecology of a mountain region in the central
    Appalachians: U.S. Geol. Survey Prof. Paper 347, 66p.

Helley, E.J., and LaMarche, V.C., 1973, Historical flood
    information for northern California from geologic and
    botanical evidence: U.S. Geol. Survey Prof. Paper 485-
    E, 16p.

Hoyt, W.G., and Langbein, W.B., 1955, Floods: Princeton
Univ. Press, Princeton, New Jersey, 469p.

Jahns, R.H., 1947, Geologic features of the Connecticut
Valley, Massachusetts as related to recent floods:
U.S. Geol. Survey Water Supply Paper 996, 158p.

Lundelius, E.L., Jr., 1967, Late-Pleistocene and Holocene
faunal history of central Texas: in, Martin, P.S. and
Wright, H.E., Jr., (eds.), Pleistocene extinctions, the
search for a cause: Yale Univ. Press, New Haven, p. 287-
319.

Miami Conservancy District, 1917, Storm rainfall of eastern
United States: Tech. Reports, part 5.

Moss, J.H., and Kochel, C., (in press), Noteworthy characteristics of the Hurricane Agnes flood, Conestoga
drainage basin, southeastern Pennsylvania: Jour. Geol.

Orton, R., 1966, Characteristic morphology of some large
flood producing storms in Texas - easterly waves: in,
Symposium of some aspects of storms and floods in
water planning: Texas Water Dev. Board Rept. 33,
p. 4-17.

Patton, P.C., 1977, Geomorphic criteria for estimating the
magnitude and frequency of flooding in central Texas:
unpb. Ph. D. dissertation, Univ. of Texas at Austin,
Austin, Texas, 225p.

_____, and Baker, V.R., 1976, Morphometry and floods in
small drainage basins subject to diverse hydrogeo-
morphic controls: Water Resources Res., v. 12, no. 5,
p. 941-952.

Rostvedt, J. O., and others, 1970, Summary of floods in the
United States during 1964: U.S. Geol. Survey Water
Supply Paper 1840-C, 124p.

Schapenseel, H.W., 1971, Radiocarbon dating of soils -
problems, troubles, hopes: in, Yaalon, D.H., (ed.),
Paleopedology, origin, nature and dating of paleosols:
Int. Soc. Soil Sci. and Israel Univ. Press, p. 77-88.

Schumm, S.A., 1963, A tentative classification of alluvial
river channels: U.S. Geol. Survey Circ. 447, 10p.

Sellards, E.H., 1934, Sturctural geology east of Pecos
River: in, Sellards, E.H. and Baker, C.L., (eds.), The
geology of Texas: v. II, Structural and economic
geology: Univ. Texas Bureau of Econ. Geol. Bull.,
no. 3401, 844p.

Shepherd, R.G., 1975, Geomorphic operation, evolution, and equilibria, Sandy Creek watershed, Llano region, central Texas: unpbl. Ph. D. dissertation, Univ. of Texas at Austin, Austin, Texas, 209p.

_____, and Schumm, S.A., 1974, Experimental study of river incision: Geol. Soc. America Bull., v. 85, p. 257-268.

Stewart, J.H., and LaMarche, V.C., 1967, Erosion and deposition produced by the flood of December 1964, on Coffee Creek, Trinity County, California: U.S. Geol. Survey Prof. Paper 422-K, 22p.

Tinkler, K.J., 1971, Active valley meanders in south-central Texas and their wider implications: Geol. Soc. America Bull., v. 82, p. 1783-1800.

U.S. Soil Conservation Service, 1954, Special storm report, storm of June 26-28, 1954, Johnson Creek watershed, tributary to the Devils River, Texas: Temple, Texas, Soil Cons. Service, 10p.

Williams, G.P., and Guy, H.P., 1973, Erosional and depositional aspects of Hurricane Camille in Virginia, 1969: U.S. Geol. Survey Prof. Paper 804, 80p.

Wolman, M.G., and Leopold, L.B., 1957, River flood plains - some observations on their formation: U.S. Geol. Survey Prof. Paper 282-C, p. 87-109.

_____, and Miller, J.C., 1960, Magnitude and frequency of forces in geomorphic processes: Jour. Geol., v. 68, p. 54-74.

VIGIL NETWORK ESTABLISHMENT FOR MONTANA'S COAL REGION

Robert R. Curry
Department of Geology
University of Montana

ABSTRACT

The impact of proposed coal-related development on watersheds of southeastern Montana is being assessed using surveys of stream channel characteristics, micro- and macro-faunas, sediments, sediment discharges, and comparative historical channel pattern changes. A 28,500 km^2 portion of the 180,000 km^2 Yellowstone River basin was intensively studied. In this area, coal extraction is expected to reach about 100 million metric tons per year by 1980 and the region is subject to rapid development of new towns, construction of railroads and highways, opening of large strip mines, and construction of coal-fired electric generation plants with ancillary water and electric transmission and diversion facilities. The previous dominant use of the region has been for cattle ranching with minor logging and farming.

Twenty-two detailed vigil network stations were established in 1975 on the headwater tributaries (2 km^2 to about 2000 km^2 drainage areas) of the Yellowstone River in the areas to be most heavily impacted. Since major stream diversions and possible dams are also anticipated on trunk streams, additional grab sediment samples, Helley-Smith bedload transport samples, and channel geometry data were taken for the Yellowstone and its major trunk tributaries as well as for the headwater areas. Together with additional stations established in 1973 and with the U. S. Geological Survey gaging station channel cross secitons, there now exists an adaptable network of data sensitive to changes in channel shape and pattern, sediment discharge, stream faunas, stream-side vegetation and aerial and ground-based photographic aspect. These data will permit assessment of watershed changes induced by changes in land-use, urbanization, compaction, flow-diversion, weather modification, and long-term secular climatic change.

Preliminary assessment of the existing data base indicates that streams such as the Bighorn and main stem of the Yellowstone that drain glaciated areas are not today in equilibrium. Over the last century, these watercourses have tended to increase meander amplitudes and change from braided to more meandering patterns. Non-glacial streams that directly drain the strippable coal-bearing areas are today very sinuous and more stable. These latter streams arise in regions of only modest relief and low (40-60 cm/annum) precipitation with high evaporation and porous sedimentary bedrock. These streams have very low volumes of runoff per unit drainage area. Flood-flow volumes are only poorly related to drainage basin area and flow volumes frequently decrease downstream. These non-glacial tributary coal region streams appear to be much more susceptible to induced changes in channel patterns and flow characteristics brought on by changes in land-use factors that affect flood routing, aquifer behavior, and sediment production. Such changes are all anticipated with large-scale coal development such that decreased agricultural uses of bottom lands can be expected. Rates of expected changes cannot be assessed before resurveys are conducted. Existing agricultural practices such as construction of stock tank dams in headwater gullies complicate assessment of causal relationships.

WEATHERING OF CALICHE IN SOUTHERN NEVADA

Laurence H. Lattman
Department of Geology and Geophysics
University of Utah

ABSTRACT

Caliche in the arid climate of southern Nevada is being weathered by mechanical breakup and by solution. Mechanical breakup is the dominant process operating on indurated layers such as petrocalcic and laminar horizons. Solution attacks the softer, non-indurated caliche layers. These softer layers may become case-hardened on exposure, after which mechanical breakup is significant.

Mechanical breakup is probably due mostly to freeze and thaw of water at all altitudes in southern Nevada. The resultant product is a caliche rubble layer which is resistant to further weathering and erosion.

INTRODUCTION

A considerable literature exists on secondary calcium carbonate deposition (caliche) in arid and semi-arid regions (Reeves, 1976). Far less effort, however, has been expended on studies of the weathering and destruction of caliche. The latter subject plays an important role in the effect of caliche horizons on landscape evolution.

Solution and mechanical breakup are the two major processes causing weathering of caliche under arid conditions. Although they generally operate together, one may be dominant in a single locality at any particular time. Which of the two processes dominates depends primarily on the type of caliche horizon. The soft calcium carbonate layers are destroyed essentially by solution and the hard layers, such as petrocalcic and laminar horizons, are destroyed by dominantly mechanical breakup.

TYPES OF CALICHE HORIZONS

The various types of caliche horizons have been described by Gile (1966) from the pedogenic viewpoint, and Lattman (1973) has discussed them from the geologic view.

Briefly, pedogenic calcium carbonate deposition develops through a set of stages starting with early incomplete nodules and stringers in the soil and develops into a calcic horizon having "a calcium carbonate equivalent content of more than 15 percent...." (Soil Survey Staff, 1967, p. 34). Such a horizon is generally soft or punky. At a later stage, if the calcium carbonate is continuous in the soil, the layer is designated a K horizon (Gile and others, 1965). If the K horizon is indurated (does not slake in water) it is called a petrocalcic horizon. K horizons are quite hard and of low permeability. Laminar layers are similar to petrocalcic horizons but are thinly laminated and generally not as thick. Dense, indurated layers similar to petrocalcic horizons may develop very locally in gully bottoms (Lattman, 1973) and are not of pedogenic origin.

One additional process affects the interaction of caliche with the weathering processes. Soft calcium carbonate layers may undergo case-hardening upon exposure and thus become indurated to a depth of up to 50 cm. Case-hardening occurs on horizontal and vertical exposures and is very rapid, developing to a noticeable extent in a few months (Lattman and Simonberg, 1971). Once case-hardened, soft calcic horizons may react to weathering in a fashion similar to petrocalcic horizons.

WATER AND TEMPERATURE RELATIONSHIPS IN A CALICHE PROFILE

About ten kilometers south of Las Vegas, Nevada is a large alluvial fan extending northward from the McCollough Mountains. The fan has a well-developed caliche profile including a petrocalcic horizon (Figure 1). Overlying the petrocalcic horizon is about 8 cm of caliche rubble mixed

Figure 1. *Caliche profile on the McCollough fan. The boulder-free layer above the hammer handle is a petrocalcic horizon. On top of the petrocalcic horizon is a layer of caliche rubble. Andesite boulders litter the surface.*

Figure 2. *Test site on the McCollough fan, showing thermister leads. The surface is composed of andesite boulders and caliche rubble.*

with quartz sand. Scattered basalt and andesite boulders litter the surface. A multi-lead temperature recorder was installed on the fan during July, 1973 (Figure 2). Five thermister leads were used and were read every fifteen minutes for several days (the five leads were all read within one second). One thermister was installed on the south side of an andesite boulder and the others at depths of 2, 4.5, 6 and 8 cm. The 8 cm-deep thermister was on top of the petrocalcic horizon. Figure 3 shows the temperature record for the 24 hour period from sunrise on July 13 to sunrise on July 14, 1973. At about 1:00 p.m. (1300 hours) water was poured on the test area (one square meter) from a sprinkler can at the rate of 2 liters per minute for 3 minutes.

Referring to Figure 3 it is seen that at sunrise on July 13 the lower levels (6 and 8 cm) were warmer than the upper levels (2 and 4.5 cm) and the andesite boulder. The boulder and the 2 cm-deep level heated up quite rapidly and by 10:30 a.m. (1030 hours) the temperature was uniformly warmer closer to the surface. By about 1300 hours the boulder and the 2 cm-deep level reached their maximum temperature (about 62°C) and held this till about 5:00 p.m. (1700 hours). The lower levels lagged behind the upper levels and reached a maximum of between 55° and 60°C at about 4:00 p.m. (1600 hours). All levels started to cool at about 5:00 p.m. and by about 6:00 p.m. (1800 hours) the upper levels were cooler than the lower levels. Several other observations may be made from Figure 3. The andesite boulder and the 2 cm-deep level kept very close to the same temperature but the levels at 4.5 cm and below were 5° to 8° cooler and within about 2°C of each other during the hottest part of the day. During the early afternoon the sky was covered to about 40 percent with isolated cumulus clouds. Shadows of these clouds caused about a 2°C temperature variation in the boulder surface and 2 cm-deep level but did not affect the lower levels.

The water percolating downward caused a sharp spike (about 8°C increase) in the temperature at 4.5 and 6 cm

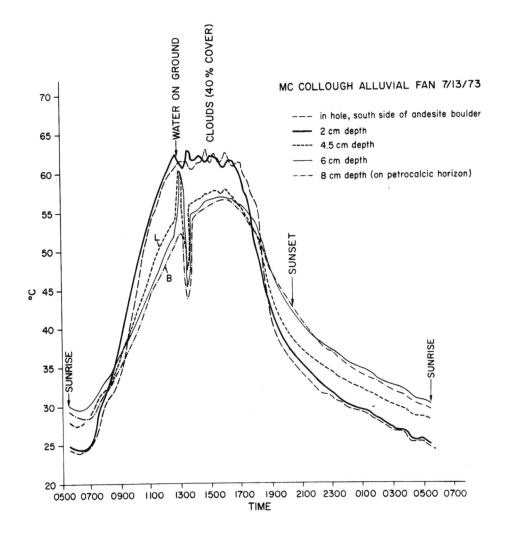

Figure 3. Temperature record in caliche profile on McCollough fan.

depths but did not appear to affect the top of the petrocalcic horizon at 8 cm. This temperature increase was followed immediately by a 10° to 12°C temperature drop at 4.5, 6 and 8 cm depth and at 1400 hours (one hour after water was sprinkled on the surface) the temperature at all levels had returned to the projections of their pre-watering traces. No water penetrated the petrocalcic horizon. The maximum range in heating and cooling, 60°C to 44°C, occurred at 6 cm depth. Observation of the profile between 4.5 and 8 cm depth (L and B of Figure 3) revealed small, scattered fresh calcium carbonate deposits in this range all over the McCollough fan. Other similar tests elsewhere on the fan gave essentially similar results even when water addition rates were increased to five liters per minute for five minutes.

From these studies it appears that petrocalcic (and laminar) layers are probably not subject to solution attack in the summer if buried and, being highly impermeable, cause deposition of calcium carbonate above them. It is also suggested that evaporation is more significant than temperature change in causing calcium carbonate deposition within a soil, because the downward percolating water is warm (from the hot surface) and hence would hold less calcium carbonate as bicarbonate. When the water cools by evaporation and could hold more bicarbonate in solution there is less water. A considerable rainfall would be required to saturate the soil above the petrocalcic horizon and flow down slope to remove calcium carbonate from the fan. Lesser rainfall would simply cause local movement and redeposition of the calcium carbonate within the soil above the petrocalcic horizon.

WEATHERING OF PETROCALCIC AND LAMINAR LAYERS

The indurated petrocalcic and laminar layers weather predominantly by mechanical breakup when exposed at the surface. The resultant caliche rubble litters the surface above the hard layer (Figure 4A, 4B) and is composed of hard, angular fragments from a few millimeters to 20 cm in major axis. The rubble shows very little or no solution effect. Caliche rubble layers in southern Nevada are up to 35 cm thick and are in sharp contact with the unbroken indurated layer beneath (Figure 4A). The rubble is best developed where indurated layers are crudely bedded (Figure 4B) and jointed.

Freeze and thaw of water is believed to cause the mechanical breakup of indurated caliche layers in southern Nevada. Data from the Corn Springs Desert National Wildlife Range near Las Vegas, Nevada indicate maximum precipitation occurs between September and April and that the ground temperature on the valley floor fluctuates above and below freezing about 35 times from December through February. The widespread, but light, winter precipitation combined with frequent freezing and thawing are the most obvious explanation of the origin of the caliche rubble. The data given above are for the valley floor and both precipitation and freeze-thaw cycles should increase at higher elevations on the fans. The sharp, but wavy contact between the rubble layer and the underlying indurated layer (Figure 4A) may be due to the irregular permeability of the rubble layer with resultant irregular depth of water penetration. As noted above, solution effects on the indurated layer (and rubble layer) are negligible.

As the rubble layer is permeable, it is resistant to gullying and such layers cover large areas underlain by indurated calcium carbonate horizons throughout southern Nevada and adjacent areas. The rubble as well as the underlying indurated layer is thus a resistant rock and small buttes around Las Vegas are capped by rubble with or without a subjacent petrocalcic horizon.

WEATHERING OF NON-INDURATED CALCIUM CARBONATE LAYERS

The softer calcic horizons, some of which cement gravel and others of which are free of coarse detritus show solution effects. The solution is apparent in a fretted

Figure 4. Caliche Rubble.

Figure 4A. Caliche rubble at surface (upper left) lying in sharp contact on underlying petrocalcic horizon.

Figure 4B. Angular caliche rubble littering surface above bedded and jointed, hard caliche layer.

223

appearance of the surface and is particularly obvious at the edges of included boulders. Areas where deep and extensive solution effects are obvious are rare in southern Nevada, apparently because of case-hardening. When calcic horizons are exposed to water at the surface they become indurated to various depths (Figure 5A, 5B). Such layers remain soft beneath the indurated crust (Lattman and Simonberg, 1971). Once case-hardened, such layers in vertical and horizontal exposures are more affected by mechanical breakup than by solution. The case-hardened surface is usually dark grey due to organic material and is rough. A thin, discontinuous rubble forms on the case-hardened surface (Figure 5B). The rubble itself frequently shows solution rounding at the edges.

CONCLUSIONS

It appears that weathering of caliche in arid areas such as southern Nevada does not always result in simple weakening of the "rock." Indurated horizons in a dry climate breakup mechanically and the resultant caliche rubble is itself a resistant rock capping buttes. Non-indurated calcic horizons undergo some solution but are also case-hardened resulting in surface induration. Subsequent to case-hardening such layers weather dominantly by mechanical breakup to yield rubble. Hence, caliche once formed will remain a resistant layer and perhaps a caprock despite weathering unless the climate becomes moist enough to remove it by solution.

Figure 5. Case-hardening.

Figure 5A. Case-hardening of several caliche layers in a
vertical exposure in a small wash.

Figure 5B. Horizontal case-hardened surface on caliche. Note
scattered rubble. Ruler is 15 cm long.

REFERENCES CITED

Gile, L.H., Peterson, F.F., and Grossman, R.B., 1965, The
 K horizon: A master soil horizon of carbonate accumulation: Soil Sci., v. 99, p. 74-82.

_____, 1966, Morphological and genetic sequences of carbon-
 ate accumulation in desert soils: Soil Sci., v. 101,
 p. 347-360.

Lattman, L.H., 1973, Calcium carbonate cementation of alluvial fans in southern Nevada: Geol. Soc. America Bull.,
 v. 84, p. 3013-3028.

_____ and Simonberg, E.M., 1971, Case-hardening of carbonate
 alluvium and colluvium, Spring Mountains, Nevada: Jour.
 Sed. Petrology, v. 41, no. 1, p. 274-281.

Reeves, C.C., Jr., 1976, Caliche - origin, classification,
 morphology and uses: Lubbock, Texas, Estacado Books,
 233p.

Soil Survey Staff, 1967, Supplement to soil classication
 system (7th approximation): Soil Cons. Service, U. S.
 Dept. Agriculture, 207p.

YARDANGS

John F. McCauley
Maurice J. Grolier
Carol S. Breed

U.S. Geological Survey
Flagstaff, Arizona

ABSTRACT

Yardangs are perhaps, the most neglected and misunderstood of the Earth's landforms. The existence of large fields of yardangs ranging from meters to kilometers in length is not well known to American geologists. In the United States many early workers overstated the role of the wind in landscape development, leaving a legacy of uncertainty about the effectiveness of the wind as an agent of surface sculpture.

Yardangs were first described along the eastern edge of the Takla Makin Desert by the explorer Sven Hedin in 1903. The term is considered applicable to all positive, streamlined, aerodynamically shaped hills regardless of size and type of bedrock in which they form (Yardang is the ablative form of the Turkestani word *yar* which means ridge or steep bank - the ablative in this case expresses the sense of removal - an appropriate connotation for wind erosion features). Wind sculptured hills have been described, often under different names, in the desert regions of every continent except Australia. They are restricted to those parts of deserts that are extremely arid where plant cover and soil development are minimal and strong unidirectional winds occur throughout much of the year. Major localities are known in Iran, Egypt, Arabia, Chad, Namibia and Peru. Scanty data suggests that they also occur in Afghanistan, Libya, and Mauritania. Minor localities have been described at Rogers and Coyote Lakes, California.

The somewhat imperfect yardangs of the Talara region of northern Peru have been known for more than fifty years.

This region is subject to torrential heavy rains about every
20 years so that these windforms bear the imprint of episodic
running water as do many of the African and Iranian examples.
The coastal desert of central and southern Peru is essentially rainless (less than 10 mm of precipitation reported at
Pisco per year and this is in the form of condensate from the
coastal fog). In the Ica Valley region of south central Peru
there are thousands of small to large totally ungullied hills
oriented parallel to one another. They occur in clusters
with individual hills showing a high degree of streamlining
in both plan and profile. These wind erosion forms occur
in the upper Tertiary Pisco Formation which consists of
white to yellow sandstones, siltstones, bentonites and thin
layers of conglomerate. In other deserts yardangs are known
to occur in Paleozoic and Mesozoic sediments and also even
in crystalline rocks such as gneisses and granites.

There is much confusion in the existing literature on
the origin of yardangs. Although abrasion and deflation
have long been considered integral parts of the wind erosion
regime, many authors have stated that abrasion alone is
responsible for yardangs. New field and experimental modeling data suggests that deflation is far more important in
yardang development than previously realized.

INTRODUCTION

The existence of large swarms of wind-sculptured,
streamlined hills of meter to kilometer size is not well
known to American geologists. Some early workers in the
southwestern United States (Keyes, 1909, among others) overstated the role of the wind in landscape development, leaving a legacy of doubt about the effectiveness of the wind as
an agent of surface sculpture.

In the mostly semi-arid southwestern United States
fluvial action dominates the landscape. In more arid regions, on the other hand, Peel (1970) described a variety of
desert landforms that are attributed to wind effects and
points out that no general studies of wind erosion features

have been done. Recent study of wind erosion features in Peru and other desert regions indicates that aerodynamically shaped erosion landforms or yardangs are more common than generally realized (McCauley and others, 1977a). This report is a condensation of that larger informal document prepared under NASA auspices for application to studies of Mars.

Yardangs were first·described along the eastern edge of the Taklimakan Desert of Chinese Turkestan by the Swedish explorer Sven Hedin (1903). The word yardang is the ablative form of the Turkestani word *yar*, which means ridge or steep bank (the ablative case expresses the sense of removal). The etymology of the word is, therefore, appropriate for positive wind produced streamlined features regardless of size and the type of bedrock in which they occur. The original definition of yardang by Hedin applied to the positive form, but the term has been subsequently misapplied by some authors (Tricart and Cailleux, 1969, p. 307-309) to the hollows or troughs between individual yardangs.

Since Hedin, streamlined, wind-sculptured erosion features have been described in many extremely arid desert regions where plant cover and soil development are minimal and where strong unidirectional winds occur throughout much of the year. Only the best described yardang localities outside of Peru will be reviewed here. Other localities where the data are scanty but where yardangs are known to occur include Egypt, Arabia, Libya, Afghanistan, and the Namib desert of southern Africa. In the United States yardangs were described by Blackwelder (1934) on the northeast side of Rogers Lake, California.

YARDANGS OF THE TAKLIMAKAN DESERT, CHINA

Yardangs of the Taklimakan Desert in the Tarim Basin, northwestern China are much smaller than those of the other desert regions we have studied. They are significant, however, because they represent to our knowledge the first described, aerodynamically shaped, wind erosion forms of positive relief.

The Tarim Basin is surrounded on the north, west, and south by mountain chains that rise up to more than 7000 m. The Lop Nur depression to the east, where the yardangs occur, is bordered on its eastern side by low mountains, on the northeast by the Kuruktag Shan, and on the south by the Altun Shan. The climate is extremely arid and characteristic of the intracontinental deserts in the temperate zone of Asia. In the winter, cold northeasterly winds prevail; in the summer the area becomes a thermal basin and the northeasterly winds abate.

Hedin (1905, p. 63, 65) described the desert floor in the eastern Taklimakan Desert as wind-furrowed with the yardangs and intervening gullies composed of hard, lacustrine clay. He surmised that the fine, yellow, fossiliferous clay once formed the bottom of a larger Lop Nur. The clay consists of several layers which vary in hardness and generally dip 2-3 degrees in a downwind direction to the southwest and west. Bulging concretions also occasionally occur in the yardang-forming unit.

Hedin (1903, v. 1, p. 350) first observed streamlined hills about 75 km west of Lop Nur in 1900, on the former delta of the Tarim River exposed within the dry bed of the Kuruk River (also known as the Kum Darya, Karuk Darya, and K'ung-ch'ueh Ho):

> we came to some exceedingly difficult country,
> namely a perfect labyrinth of clay 'terraces'
> with sharp-cut edges, which the natives called
> yardang; and as this is a very graphic descrip-
> tive word, I shall for the future use it when
> speaking of these scarped formations.

The most common configuration of the yardangs in the Lop Nur region (Hedin, 1905, p. 65-66) is a flat-topped ridge, with ragged edges as shown in Figure 1, but sharp-topped ridges are also common as shown in Figure 2. Some ridges are abruptly broken off on the northeast (upwind side) and gently slope downwind to the southwest. Undercutting of the lateral sides of yardangs by the wind is common, and so are slump blocks on the gully floors nearby.

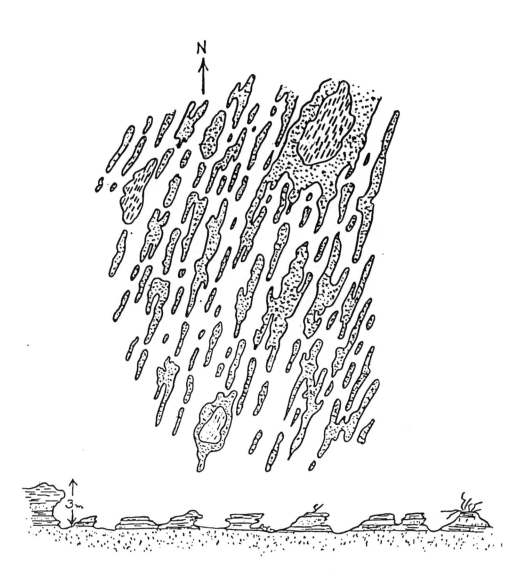

Figure 1. Yardangs of the Lop Nur region, sketched from Hedin (1905, fig. 138). Yardangs range in height from 20 m (shown in plan view) to 2 to 3 m (shown in profile view) and are commonly hundreds of meters long.

Small yardangs as much as 2 m high, decreasing to only 0.3 m southeastward, were described (Hedin, 1905, p. 58) near the dry bed of the Kuruk River. Larger yardangs were observed by Hedin during an earlier trip to the east end of the Taklimakan, presumably on the northeast side of the old lake

Figure 2. Sharp-topped yardangs near Lop Nur, sketched from Hedin (1905, pl. 27).

bed.   There the yardangs consist of clay ridges that are about 6 m high and 9-12 m across the top (Hedin, 1903, v. 2, p. 94).

YARDANGS OF IRAN

Among the largest and best-described yardangs are those of the northwestern part of the Lut Desert in Iran.   They occur in an area about 150 km long and 50 km wide in the Lut depression, approximately 60 km east of Shadad (Figure 3). According to Gabriel (1938, p. 197) the Lut is the lowest and hottest part of the Persian desert.   Like Lop Nur, it is surrounded by mountain ranges that locally rise to heights of more then 3000 m above the desert floor.   It is a region of intense wind erosion, dominated in the summer by hot dry winds from the north, and in the winter by the *Bad kessif* or dirty wind from the south, which carries enormous quantities of dust and sand, completely obscuring visibility. The region

238

Figure 3. *Landsat picture (21 January 1973) showing yardangs of the Lut Desert, Iran. Inset shows Mariner 9 frame (DAS06823253) of apparent wind scour features in the Amazonis region of Mars at the same scale as the Landsat picture.*

receives less than 50 mm annual rainfall.

The Lut Formation in which the yardangs of Iran are formed consists of fine-grained, horizontally bedded silty clay and limy, gypsiferous sand. The estimated thickness of the formation varies from about 135-200 m. The unit is almost everywhere encrusted with a mixture of salt, gypsum, and silty clay. Its grain size, stratigraphy (including salty beds) and its coarsening toward the margin of the

basin suggest deposition in a playa-type environment
(Krinsley, 1970).

The Lut yardangs are about an order of magnitude larger
than those of northwestern China, and were first described
by Gabriel (1938, p. 197) as:

> deposits of the former Lut lake, which the ele-
> mental forces of the desert have chiseled into
> shapes having a fantastic resemblance to build-
> ings so that they are called by the Iranians
> *Shahr Lut*, desert cities and by the Baluchis
> *Kalut*, desert villages.

The *kalut* (yardangs) trend NNW-SSE parallel to the
direction of the prevailing winds and are separated by
"boulevards" a hundred meters or more wide. In plan view,
individual hills show a high degree of streamlining, with
smaller wind-scour features superimposed on larger ones.
Intense gullying on the hillsides indicates that running
water, however episodic, competes actively with the wind for
dominance of the topography. The gullies are, however, sub-
ordinate features superimposed on the yardangs, as shown in
Figure 4. The interplay between these two geologic processes
gives the yardangs their curious similarity to desert ruins
when viewed from the ground.

Krinsley (1970) suggests that gullying of the yardangs
may have taken place episodically throughout the Pleistocene
until now. Gullies may be seasonal features, active during
the wetter winter seasons, with wind erosion dominating
during the other seasons. Regardless of which of these
alternatives is correct, most authors agree that the yardangs are primarily windforms that have subsequently been
modified and locally disfigured by the effects of running
water. Thus, the wind is the dominant agent of land sculpture in the Lut Desert. The action of water has not been
powerful enough to destroy the regional symmetry, the characteristic streamlined shape and convex profiles of these
yardangs, although it has contributed widely to their
ruiniform appearance.

Figure 4. Oblique aerial photograph by D.A. Krinsley of typical Lut yardangs showing flank gullies.

YARDANGS OF AFRICA

Erosional windforms have been recognized in North Africa and the Arabian Peninsula since the dawn of history, but there are surprisingly few detailed accounts of these features. Dune fields as well as the regional geology and archeology of these regions have been the principal attractions for most of the early explorers and geologists. The search for both long- and short-term climatic changes also has received more attention than the geomorphology.

Many yardang localities are known but except for the work of Mainguet (1972) in the Borkou region of Chad between the Ennedi and Tibesti Mountains, the yardangs are described so incompletely that it is difficult to establish any context for their distribution or to know whether they are fossil forms carved during dryer and windier periods, or

241

products of the current wind regime. In addition, erosional windforms have been given a confusing variety of local names and interpretations that further blur our understanding of the yardangs in this part of the world.

In Egypt, medium-size landforms sculptured by the wind have long been known, but they are very poorly described. Some of these aerodynamic landforms are an important part of the human history of this region. The Sphinx of Egypt (about 15 km southwest of Cairo), portrays the familiar head of Ramses II upwind from the recumbent body of a lion. It was carved out of an already streamlined hillock of Mokottam limestone, a nummulitic limestone of Eocene age, about 4500 years ago.

Sphinx-like hills cut in old lake beds between El Kharga Oasis and Gilf Kebir were described as "mud-lions" by Bagnold (1939, p. 283). Walther (1924, p. 207) was one of the first modern geomorphologists to associate sphinx-like hills with wind erosion. Unlike earlier workers, Walther (1924) emphasized the importance of deflation in the formation of these features.

Wind erosion forms in the Sahara, west and southwest of the Libyan Desert have been documented only in the last twenty years. In south-central Algeria, mounds sculptured by the wind out of gypsum-bearing clay of early Cretaceous age and upper Cambrian claystone have been described near In-Salah (27°12'N, 2°30'E) by Capot-Rey (1957, p. 242-246). In Arabic, these mounds, which are 2 to 5 m high, are called *Zbara* (sing). They are elongated in a northeast-southwest direction, parallel to the dominant wind, have a blunt rounded face upwind, and a crestline tapering downward.

South of the Tibesti Mountains, the Borkou region in northern Chad shows an abundance of wind-eroded landforms of gigantic dimensions. The circum-Tibesti region appears to have one of the most regular wind regimes in the Sahara, being influenced by strong northeast trade winds from September through May, and the Borkou region, because of its

242

position between the Ennedi Mountains on the southeast and the Tibesti Mountains to the north, is one of the windiest localities in Chad. The winds are controlled by the regional topography and are very constant in direction. Eight out of twelve months a year they blow from the north-northwest with average velocities of about 3 m/sec during the warm season and 22 m/sec during the cold months.

The most complete descriptions of wind erosion features in the Borkou region are by Mainguet (1968, 1970, 1972, 1974). Klitzsch (1966), Hagedorn (1968), and Hagedorn and Pachur (1971) have also described the Tibesti region and its various windforms. Worral (1974) summarized some of the work carried out by others, and amplified it with additional observations.

Landsat pictures show the largest streamlined windforms present in the Paleozoic, Mesozoic, and even the Precambrian rocks of the circum-Tibesti region (Figure 5). Huge yardangs observed west of Ounianga Kebir are commonly more than 20 km long and 1 km or more wide. They are separated by troughs ranging in width from 500 m - 2 km.

Undoubtedly there are many other yardang localities in western Asia and North Africa. Many of these are probably well known to oil company geologists who have worked in Libya and Algeria, but they have received surprisingly little attention in the literature. Capot-Rey (1957) reports that the coast of Mauritania and Tidikelt is notable for its constant winds, almost comparable to the wind regime of the Borkou area, and an inspection of Landsat pictures of the area suggests that yardangs may also be present there.

Late in the course of this study, we became aware that yardangs also occur in the Namib Desert of Namibia (South West Africa). The Namib is a cool coastal desert like that of Peru that is also nearly rainless and almost devoid of any vegetation.

The earliest and most dramatic account of the wind erosion forms of the Namib Desert was provided by

243

Figure 5. Landsat picture (30 October 1972) showing very large yardangs in the Paleozoic rocks of the Borkou region, Chad. Unusually large single yardangs 20 to 30 km in length are seen.

Harger (1914) who described a variety of small and fantasti-
cally shaped wind corrosion forms along with pits and hollows that are now generally referred to as tafoni. He also
recognized that great valleys have been carved out of the
limestones and other bedrock in the Pomona region by wind
action. The hills between these valleys are almost surely
yardangs, but Harger did not use the term.

Kaiser (1926, v. 2, p. 214-232) discussed wind erosion
effects in great detail in a monumental work on the region
between the Orange River and Lüderitz. Several of Kaiser's

aerial photographs suggest bedrock streamlining. The excellent geologic maps at the 1:25,000 scale included in his
report show numerous, apparently streamlined outcrop patterns that range in size from a hundred meters or so in
length up to tens of kilometers. Figure 6 is a relatively
recent photograph that provides further evidence that large
yardangs are indeed present in the Namib Desert.

YARDANGS OF NORTH AMERICA

Yardangs have been reported from only a few localities
in the western United States. Blackwelder (1934, p. 160)
mentioned the presence of yardangs both east and west of
Tonopah, north of Silver Peak and northeast of Carson Sink
in Nevada, but did not describe them. Yardangs also have
been reported from Coyote Lake, about 27 km northeast of
Barstow, California (Hagar, 1966). Long, narrow, undrained

*Figure 6. Low-altitude, oblique aerial view of probable yardangs
in Precambrian rocks of the Namib Desert southeast of
Lüderitz (photograph by H.T.U. Smith).*

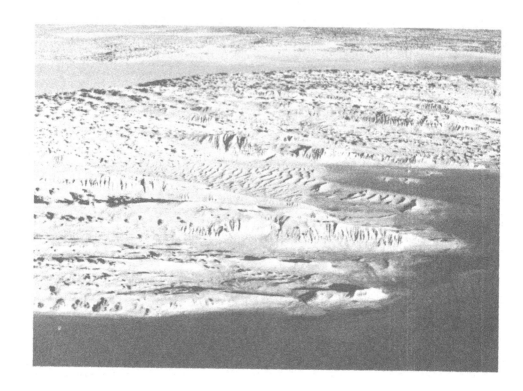

Figure 7. Aerial photograph of yardangs in the southern end of the Rogers Lake field.

grooves or trenches excavated in bentonitic clays in the South Dakota badlands, between Conota Station and Cuny Table, have been called yardangs by Baker (1951). The only North American locality described in detail (Blackwelder, 1934) is on the northeast side of Rogers Lake, California, at about 34°58'N, 117°47'W (Figure 7).

Rogers Lake is in the western part of the Mohave Desert, where the ingress of marine air masses is blocked by coastal ranges, and annual precipitation is less than 12.5 cm. A single storm may yield all of the rainfall for one locality for a month or even a year. Strong winds from the west are typical throughout the year (Thompson, 1929). The yardang locality now lies within the confines of Edwards Air Force Base and was visited briefly by the authors in December 1975 and again in 1977.

The yardangs, as described by Blackwelder (1934, p. 160), occur in a broad belt of dune sand that marks the old

shoreline of Pleistocene Rogers Lake. The sand is coarse
and mingled with clay pellets and clay laminae which are
more resistant to wind erosion than the more friable quartz
sands. The dune sands at Rogers Lake are cut into a series
of sharp ridges and round bottomed chutes with a total relief of about 7.5 m. The bottoms of the chutes are covered
locally by coarse sand to granule ripples.

Blackwelder speculated at some length on the mode of
formation of yardangs. He ascribed the origin of the
troughs to possible inequities in the material itself, its
structure or its original surface. In the latter case,
Blackwelder appealed to the presence of occasional rills,
which might have provided the wind with pathways necessary
for the excavation of troughs.

Our brief reconnaissance suggests strongly that the
simple abrasion model for yardangs presented by Blackwelder
is incorrect, and that deflation is far more important than
he allowed. The abrasion model of Blackwelder cannot explain the overall streamlined shape, nor can it explain the
sharp but smooth ridge crests of the yardangs. Abrasion
effects, however do contribute to some undercutting at the
bows and flanks and to lowering of the troughs. The major
process by which the positive features, the yardangs,
attain their aerodynamic shape is thought to be deflation.
The winnowed appearance of the rocks themselves, and the
lack of small scale grooving and scouring is strong evidence
for this. We believe that the quartz grains within these
weakly indurated sediments become gradually loosened by
weathering and as the fine clay matrix breaks down the en-
closing fines are swept away in suspension. Episodic
strong gusts then provide the lift necessary to dislodge the
sand into the troughs where it is accelerated by funneling
and swept away mostly in saltation. This sand is ultimately
banked up into the lees of the yardangs as seen in Figure 6.

YARDANGS OF THE COASTAL DESERT OF PERU

The coastal desert of Peru extends some 2200 km from Tacna near the Chilean border on the south to Tumbes near the Equadorian border on the north (Figure 8). It is only about 50-100 km wide, flanked by the foothills of the Andes on the east and by the Pacific Ocean on the west. The desert covers about 10 percent of the total land area of Peru and is traversed along most of its length by the Pan American Highway.

Bellido (1969) presented a summary of the geology of Peru, including a generalized geologic map, upon which some of the following description is based. The coast of Peru is one of general emergence, with marine terraces and sea stacks present up to elevations of 400 m above the present shoreline. In the Tumbes-Talara area, numerous uplifted and moderately dissected marine terraces of Tertiary age are present. This northern part of the desert is only semi-arid, but south of Piura the aridity increases and the desert reaches its widest extent over the broad flat surface of the Sechura basin. Much of the Sechura is characterized by sparse vegetation and large trains of barchans. Parts of the Sechura lie below sea level, and Smith (1963) suggested that the low-lying parts of the Sechura were formed by deflation. From Chiclayo southward to the Paracas Peninsula, the emergent coastline includes numerous steep wave-cut cliffs that range in height from a few tens of meters to a maximum of about 100 m north and south of Lima.

Between Tumbes and Trujillo isolated blocks and hills of crystalline rock of Jurassic age occur as marine platforms, transected by numerous large valleys and pampas that head in the Andes to the east. Some of the streams in these transverse valleys carry water to the ocean throughout the year, but others flow only after the rainy season which generally begins in November in the highlands. More than 50 such valleys transect the coast of Peru, and most of the population and almost all agricultural acitivity is confined to these areas.

248

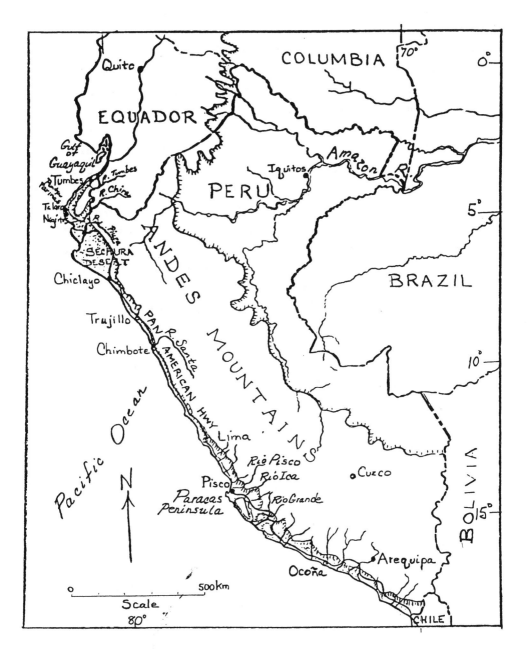

Figure 8. Index map of Peru. Stippled pattern shows the approximate extent of the coastal desert.

249

From near Trujillo on the north to the vicinity of
Ocoña on the south, the inland border of the desert consists
of intensely dissected, intrusive rocks of upper Cretaceous
to lower Tertiary age. These rocks, which rise to heights of
1000-2000 m in a belt 50-100 km wide, are generally referred
to as the "Coastal Batholith." Large alluvial fans spread
out over the coastal desert from numerous steep gorges and
ravines incised into these rocks.

From the Paracas Peninsula southward, the coast is
marked by subdued, rounded mountains whose maximum elevations are about 1500 m and by high, steep wave-cut cliffs.
Between this mountain range and the Andes is a central platform or valley which corresponds to the coastal platform of
northern Peru. This central platform is generally less than
20 km wide and lies at elevations of 500-1000 m. It slopes
gently southwestward, and is extensively covered by piedmont
sediments that are transported during flash floods from
valleys along the Andean side of the platform. This southern region is even more arid than the north coast.

The coastline of South America begins a bend to the
east near latitude 5°S and Punta Pariñas, about 12 km south
of Talara, marks the westernmost point on the continent.
The Talara region is underlain by block-faulted Tertiary to
Holocene sandstones and shales. Remarkably well-developed,
broad flat-lying Quaternary marine terraces (*tablazos*) with
an average elevation of 75 m extend inland from the coast
to the foothills of the Andes. Since the early 1900's the
Talara region has been a center of petroleum production in
Peru.

Unlike the coastal desert of central to southern Peru,
where precipitation sufficient to produce collected runoff
is exceedingly rare, the Talara region receives spasmodic
heavy rainfall from El Niño events. Bosworth (1922) reported
that severe floods were experienced in 1891, but no rain
had fallen in the area for some 30 years prior to his investigations. As a result of El Niño, the *tablazos* and the
hinterlands away from the coast are intensely gullied, and

mudflow deposits along old stream courses are common along with extensively developed badlands topography. The general aspect of the terrain is similar to that of the semi-arid regions of the world where the imprint of running water is clearly predominant. Nevertheless, erosional windforms of considerable size are present.

Wind erosion forms of the Talara region in northern Peru were originally described by Bosworth (1922), who did not use the term yardang for these features but rather called them streamlined hills, and likened them to inverted ship hulls. The windforms occur in weakly to moderately consolidated Upper Eocene to Paleocene sediments of the Lobitos and Negritos formations, consisting of gray to brownish shales and sandstones 4500-7500 m thick. These units are overlain by Quaternary alluvial, lagoonal and eolian deposits. The latter are sand sheets and various types of dunes which are moving inland from the present shoreline. Intense faulting has produced highly variable strikes, and dips of 15-20° are common in these wind-eroded beds. The strikes of the beds and the trends of minor faults locally control much of the wind furrowing.

Most of the Talara yardangs are not ideal windforms. Individual features are generally no more than a few hundred meters long and tens of meters high so that they are considerably smaller than those previously described. In addition, they are often ragged in shape as a result of flank gullying related to episodic El Niño events. Their shapes are controlled in large part by the complex geologic structure of this region and to a lesser extent by runoff. However, no large-scale, integrated drainage patterns can be recognized in the immediate vicinity of the yardangs and it is difficult to relate the forms observed to running water (Figure 9).

The area between the Paracas Peninsula and the mouth of the Rio Ica in southcentral Peru is part of a crescent-shaped structural basin that lies between the coastal Cordillera and the foothills of the Andes. It is underlain

251

Figure 9. Aerial photograph of streamlined hills south of Negritos showing gullying on flanks due to heavy rains that occurred about one year before the photograph was taken.

mostly by late Tertiary to Holocene sediments that are extensively wind-eroded from the vicinity of Ocucaje to the sea.

Principal areas where erosional windforms occur are at the north end of the neck of the Paracas Peninsula, and the region to the west and south of Pozo Santo (Cerro Lechuza) where a variety of erosional forms have been described previously by Grolier and others (1974) and McCauley and others (1977a). The lower 50-60 km of the Ica Valley at the south end of the basin contains hundreds of yardangs up to kilometers in length. Some of the yardangs of this region, although not the largest encountered in our survey, have almost perfect aerodynamic shapes.

The most magnificent of the Rio Ica yardangs occur in the Pisco Formation of upper Oligocene to upper Miocene age.

The unit consists of fine-bedded white siltstones containing
diatomites and clays along with diatomite tuffs, ash, and
gypsiferous, fine-grained sandstones (Gilboa, 1969; ONERN,
1971a, b). Thin layers of conglomerate and phosphatic
nodules are present locally. Veinlets of gypsum abound and
stand out in relief from the country rock. The unit becomes
more argillaceous and tuffaceous southward from the type
area near Pisco. Its thickness varies from 650 to 800 m.

   This part of Peru has one of the most unusual wind
regimes in the world. Dune fields and yardang complexes
curve gently through an arc of 180° and converge on the city
of Ica, which is bounded on the west by enormous star dunes
and to the east by the Andes. The crescent-shaped intermontane trough previously described and the orientation of
the coast with respect to the trade winds appear to be controlling factors. Northwesterly and southeasterly winds
prevail in the valley between Ica and Ocucaje (ONERN, 1971b,
v. 1, p. 55). The former originate at sea and occur most
frequently in the morning and late in the evening; whereas,
in the middle of the day, southeasterlies blow from the
land toward the sea. These winds result from differential
heating of the land and sea and are superposed on the regional trade winds and local winds controlled by topography.
The mean annual wind velocity at Ica (because of its shel-
tered location) is only 7 km/hr, less than half the mean
annual velocity recorded at Pisco nearer to the coast.

   The southerly orientation of this unique valley undoubtedly permits the regional trade winds and the sea
breeze to reinforce one another, to produce unusually
strong winds throughout most of the year. During our investigations of the yardangs to the west of Villacuri, we
experienced very strong afternoon winds. They peaked, as
in the Paracas region, at about 4:00 to 5:00 p.m. and were
of sufficient velocity to move coarse sand by saltation and
the granules in the ripples between the yardangs by traction.

   The region between the city of Ica and Ocucaje along
the Rio Ica is extensively irrigated and cultivated. The

Ocucaje area is one of the major centers of viticulture in
Peru. Outside the narrow cultivated part of the valley and
below Ocucaje where surface and ground water supplies are
scarce, the landscape is barren and intensely wind-sculptured.
The appearance of the middle and lower Ica Valley is almost
unworldly--no vegetation and no obvious water-cut gullies.
The skyline is broken only by what look like fleets of
inverted ships, kilometers long and composed of the brighter
buff to whitish beds of the Pisco Formation.

Figure 10 is a geomorphic sketch map of the main area of
interest between Ocucaje and the mouth of the Rio Ica. The
southern or down-valley part of the complex is marked by
long, canoe-shaped hills separated by flat-bottomed, dark,
lag-covered surfaces (Figure 11). Many hills have long upwind prows and tapering downwind tails; others show the
reverse relation, are blunt in the upwind direction and
gradually taper downwind (Figure 12). The Yardangs in the
southern part of the Cerro Yesera complex tend to be highly
elongated, closely nested and separated by straight to only
slightly sinuous yardang troughs. Isolated individual hills
show a high degree of streamlining and approach the shape
of inverted canoes. The middle part of the Cerro Yesera
field shows hybrid topography with numerous sinuous relict
watercourses that have been taken over by the prevailing
southerly wind to become yardang troughs. Former stream
segments that were oriented in the north-south direction
have been enlarged and widened; those that lay at sharp
angles to the wind are now almost obliterated. The upper
part of the Cerro Yesera yardang field shows numerous semi-
streamlined complex hills. Streamlining of individual hills
appears to be less well-developed in the north than in the
southern part of the field. Length-to-width ratios of
individual hills are only about 3:1 in the north, whereas,
ratios of 10:1 or greater are common in the southern part of
the field. Yardang troughs in the northern part of the
field are also less well-developed than those in the south,
and are not cut down to what appears to be the local base

254

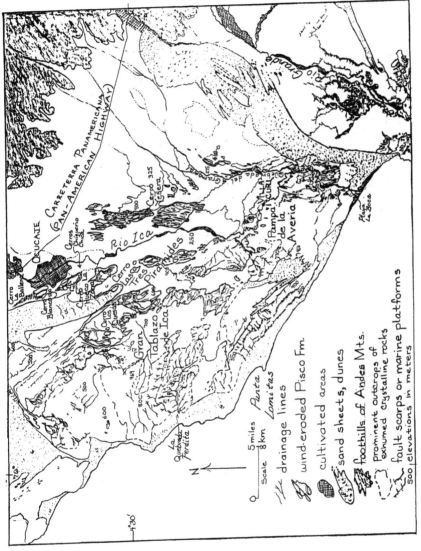

Figure 10. Geomorphic sketch map, based on aerial photographs, showing the distribution of the major landforms in the Ica Valley region, Peru.

255

Figure 11. *Oblique aerial view of the southern end of the Cerro Yesera yardang field showing kilometer-scale, well-streamlined yardangs. The largest single yardang is about 2 km long. Broad, flat-floored troughs as well as U-shaped troughs are present between yardangs.*

level (the unconformity between the Pisco Formation and the underlying rocks).

The Cerros de los Tres Piramides area, about 7 km southwest of the Cerro Yesera yardangs and on the opposite side of the river (Figure 13) shows a somewhat different type of wind erosion pattern. There the yardangs are generally more steep sided and more widely separated from one another than in the Cerro Yesera complex. Some tend to be almost triangular in shape and to have pronounced stairstep profiles. The Cerros de los Tres Piramides yardangs are surrounded by a bare rock, wind-denuded surface with almost no trace of fluvial gullying or granule ripples. Here, slight inequities in the Pisco Formation, resulting either from original

256

Figure 12. Oblique aerial view of the elongated yardangs in the
southern part of the Cerro Yesera yardang field; these
yardangs are mostly free of sand, but their bases are
surrounded by aprons of loose sand. Ripple trains on
the dark lag surface of the yardang troughs show
pattern of wind flow.

stratigraphy or from subsequent structural deformation, have
been etched into unusual and complicated patterns about a
meter or more high.

On the Pampa de la Averia farther south in the Ica
Valley, about 5 km to the west of Villacuri, the Pisco Formation has been almost completely eroded away from the underlying crystalline rocks. About 600 to 800 meters of erosion
must have occurred in this region, but how much of this
total is exclusively due to wind action is unfortunately
indeterminate. The erosional remnants of the Pisco Formation
that are present have attained smooth, almost perfectly
streamlined shapes. These features are up to 1.5 km long
and from 30-50 m high (Figure 14). Generally they are
widely separated so that individual yardangs are less

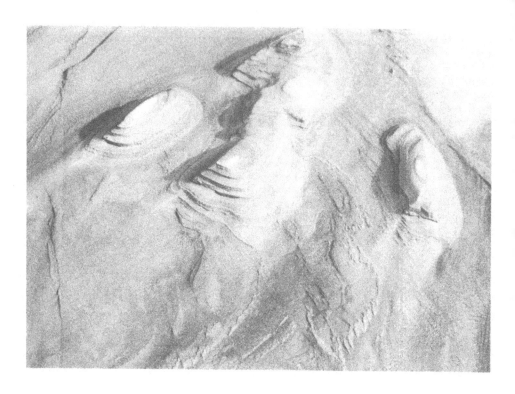

Figure 13. Oblique aerial view of the triangular-shaped yardangs of the Cerros de los Tres Piramides field, emphasizing the planimetric shape of the yardangs and their stair-step profiles.

affected by the flow patterns around upwind topography than they are in the more tightly packed fields. The topography of this region represents the most advanced stage of eolian sculpture known in Peru. The yardangs of the Pampa de la Averia are particularly notable for their smoothness and degree of aerodynamic perfection. The flow pattern seen in the surrounding ripple trains is almost identical to that observed around airfoils as well as ship hulls.

The backs of many of the yardangs in the Pampa de la Averia are quite narrow and the side slopes are about 30° (Figure 15). The surface is covered with a lag of gypsum plates and fine dust--little or no sand is present. This surface is remarkable stable even during the high afternoon winds, but any slight disturbance causes the fine gypsifer-ous dust to be blown away in suspension.

Figure 14. Oblique aerial photograph showing well-streamlined yardangs in the Pampa de la Averia, surrounded by ripple trains that diverge at the prow and converge in the downwind direction around the yardang flanks.

The yardangs of the Paracas-Ica region are locally interspersed with fields of small, meter-scale, convex-upward nubbins known locally as *quesos* (small, round but steepsideed mesas) and wind-stripped "onion skin" topography. These non-streamlined erosion features have a disquieting resemblance to some landforms that are generally explained as due to spasmodic fluvial activity in the less arid deserts. In Peru, the close association of streamlined and non-streamlined forms in exactly the same climatic regime suggests that in this region, at least, they are all of eolian origin. Figure 16 is a dramatic example of the close association of these various forms on a round, horizontally bedded, gently convex upward hill. This hill has "onion

Figure 15. Ground photograph showing sharp crest and steep (about 30°) flanks of a yardang in the Pampa de la Averia.

skin" topography at its base and superimposed on this are numerous scattered *quesos* that lie along a discrete horizon. The crest of the hill is marked by a well-streamlined yardang about one hundred meters long composed of the more massive, lighter toned beds of the uppermost part of the local sequence. The more competent Tertiary sediments of this region tend to produce "onion skin" topography. Where well-cemented zones (mega-concretions?) are present they are preferentially etched out in relief to produce *quesos*. The more massive, aerodynamically incompetent siltstones and tuffaceous units develop into streamlined yardangs or forms of minimum resistance.

Figure 16. Onion skin topography, quesos, and a well-streamlined yardang in close association on the same hill.

SUMMARY AND CONCLUSIONS

Our reconnaissance study of the wind erosion forms of Peru and literature survey of the other desert regions of the world has shown that yardangs are far more common than realized and that they span a wide range of sizes and rock types. The existing texts in geomorphology either ignore their existence or arbitrarily state that they are small inconsequential features restricted to weakly consolidated lacustrine sediments.

By priority and etymology, we believe the term yardang should be generally applied to wind-streamlined, elongate, positive topographic forms, regardless of size and type of material in which they occur as long as that material is at least weakly consolidated. We thus distinguish between yardangs which are erosional bedrock features, and the multitudinous bedforms (dunes) that are produced in non-consolidated materials by wind action. The restriction of the term yardang to streamlined or elongated positive features is in keeping with the meaning of the Turkestani word *yardang* (ridge from which materials are being removed) and Hedin's original usage.

Yardangs, regardless of size, can be recognized by their parallelism and similarity to the inverted ship hulls, oriented into the direction of the prevailing wind. They may have concave downward tapering bows or the upwind ends may be convex and bulbous. The highest and widest part of the structure is generally about one third of the way between the bow and stern in a well streamlined yardang. Where alternating competent and incompetent layers are present, local irregularities of form are common. Overhanging prows occur where competent layers overlie incompetent ones. Aerodynamic perfection can be marred by flank gullies or various types of mass wasting scars if erosional processes other than the wind episodically operate. The downwind ends of yardangs are characterized by gently tapering bedrock surfaces or elongate sand tails. Yardangs often occur in closely packed arrays or fleets separated from one another by either U-shaped troughs or flat bottomed troughs. They also occur as widely spaced features on wind-beveled surfaces of plains. These yardangs generally show a high degree of streamlining because they represent more mature forms than those in the fleets.

Yardangs seem to be restricted to sand-poor areas within the extremely arid, core regions of the Earth's deserts. These regions are generally characterized by an almost complete dearth of vegetation and extremely strong, almost unidirectional winds throughout most of the year. Yardangs probably cannot form in regions where the prevailing winds shift direction throughout the year no matter how arid these regions might be. Yardangs almost surely cannot persist long in recognizable form in regions where the climate has shifted to more humid conditions. Even a modest vegetative cover would be sufficient to check further wind erosion and the effects of running water would rapidly mask their original ship-like shapes.

There is much confusion in the existing literature on the origin of yardangs. Many authors have stated that they are exclusively the product of abrasion or sandblasting. Indeed, the early accounts of the western explorers of the Taklimakan in China did emphasize the role of abrasion in cutting the yardang troughs. Blackwelder (1934) stated that both abrasion and deflation are integral parts of the wind erosion regime but that abrasion alone was responsible for the Rogers Lake, California, yardangs. Our brief study of these features reveals that abrasion effects are minimal and that the role of deflation in forming yardangs has been understated.

Bosworth (1922, p. 295-298) was the first geologist to our knowledge to call attention to the importance of deflation in the development of yardangs. He discussed at some length how a hill can attain a streamlined shape and pointed out that an irregularly shaped obstacle offers more wind resistance than a streamlined body. The wind currents around such an irregular body would operate by erosion and deposition to produce a form that would be adapted to the local wind regime and offer minimum resistance to it. He likened this adaptation process to the mathematical problem of finding the ideal shape for a ship's hull with the conditions reversed. The ideal form for a hill in strong unidirectional winds should approximate the shape of the hull of a racing yacht turned upside down. Bosworth went further to state that the shape should vary somewhat according to the velocity of the wind (slender shapes would correspond to areas of higher speed). Finally Bosworth stated that much of the wind erosion observed takes place without the aid of sand. The rocks disintegrate and the wind simply removes the particles.

In order to better understand yardang development, a

series of preliminary qualitative experiments were conducted
(McCauley and others, 1977b). The details of this work will
be the subject of a subsequent report. These experiments confirmed, however, that irregular topographic shapes are quickly modified by unidirectional winds into streamlined or
quasi-streamlined forms. Both erosion and deposition are
important in creating these forms.

As suspected from our work in Peru and at Rogers Lake,
California, the watercourses in our topographic models that
were oriented into the prevailing wind became chutes through
which most of the sediment transport occurred. Models of
closely spaced, inverted racing ship hulls consisting of
soft over hard layers and vice versa, all of approximately
the same grain size, gave insight into yardang development.
The surface upon which the ship forms were placed was de-
flated rapidly at the beginning of the experiment and the
yardangs in these models gained in relief without much
change in form. When local base level was achieved (the
wooden board on which they were mounted) the yardangs began
to evolve in shape. Those with hard layers over soft were
modified more dramatically than the others due to undercutting, collapse, and the rapid sweeping away of the debris
that fell into the troughs. The hard-over-soft layer models
also developed sharp keels and a long, concave tapering bow
despite having been originally oriented stern first into the
wind. In contrast, the bow first, soft-over-hard layer
models developed blunt bows and broad, wide convex backs.

Grain movements were by traction, saltation, and suspension with most of the action in the troughs. Individual
grains were episodically plucked from the yardang surface
and moved into the troughs or flew out of the model area
without reimpacting the surface. As in the field, a fragile
wind-plucked surface was produced that stood out in bas-
relief on the yardangs. Abrasion effects on the "bedrock"
were minimal to non-existent in these experiments except in
the troughs where rapid downcutting occurred. Here the
higher flux of moving grains "kicks up" far more of the

264

"bedrock" than on the flanks and backs of the adjacent yardangs. Although far from conclusive and limited by obvious scaling problems, these experiments seem to confirm our field observations that deflation as well as abrasion is critical to yardang formation. Abrasion appears to be most important in the troughs and deflation on the flanks and crests. The relative efficacy of each process in the production of eolian erosion forms is still not clear. These two processes, plus traction which is responsible for moving the coarser grains, must maintain a delicate balance to produce a streamlined landform. It will be necessary to monitor the flow fields around different types of yardangs both in the field and the laboratory in order to better understand the details of the erosion process by which yardangs form. This work has begun in the laboratory and at Rogers Lake, California. Application of aerodynamic theory to the problem will also be necessary. Our descriptive work to date, however, indicates strongly that deflation or aerodynamic lifting is far more important in the production of wind erosion forms than is generally accepted by U.S. geologists at this time.

ACKNOWLEDGMENTS

Special thanks are extended to Gen. Manuel Valencia, Director-General del SAN who arranged for aerial reconnaissance studies and photography during our 1973 trip to Peru. David R. Scott, Director, NASA, Hugh L. Dryden Flight Research Center, Edwards, California, kindly arranged for our reconnaissance of the Rogers Lake yardangs. George E. Ericksen, U.S. Geological Survey, served as leader and principal organizer of both our expeditions to Peru. Daniel B. Krinsley, U.S. Geological Survey, kindly reviewed the section of this report on the yardangs of the Lut Desert, Iran. Maurice J. Terman, U.S. Geological Survey, graciously made available unpublished data on the yardang occurrences in China.

The work was sponsored by the National Aeronautics and Space Administration under contract W-12, 872, Planetology Programs, Office of Space Science.

REFERENCES CITED

Bagnold, R.A., 1939, An expedition to the Gilf Kebir and Uweinat, 1938: Geog. Jour., v. 93, p. 281-313.

Baker, C.L., 1951, Yardangs: (abs.) Geol. Soc. America Bull., v. 62, p. 1532.

Bellido, B.E., 1969, Sinópsis de la geológia del Perú: Perú, Servicio Geol., Min., Bol. 22, 54p.

Blackwelder, Eliot, 1934, Yardangs: Geol. Soc. America Bull., v. 45, p. 159-166.

Bosworth, T.O., 1922, Desert conditions and processes in the desert of Tumbes, Peru: in, Bosworth, T.O., Geology of the Tertiarty and Quaternary periods in the northwest part of Peru: Macmillan and Co., London, p. 269-309.

Capot-Rey, Robert, 1957, Sur une forme d'érosion éolinne dans le Sahara Francais: Koninkl. Nederlandch Aardrift-skundig Genootschap Tidjschrift, 2d ser., v. 74, p. 242-247.

Gabriel, Alfons, 1938, The southern Lut and Iranian Baluchistan: Geog. Jour., v. 92, p. 193-210.

Gilboa, Yaakov, 1969, The groundwater geology in the rain-less coastal area of Peru: unpbl. Ph. D. Thesis, Hebrew Univ., Jerusalem, 149p.

Grolier, M.J., Ericksen, G.E., McCauley, J.F., and Morris, E.C., 1974, The desert landforms of Peru; a prelimin-ary photographic atlas: U.S. Geol. Survey Interagency Rept. Astrogeology 57, 146p.

Hagar, D.J., 1966, Geomorphology of Coyote Valley, San Bernardino Valley, California: Unpbl. Ph.D. Thesis, Univ. Mass., 210p.

Hagedorn, Horst, 1968, Über äolische abtragung und formung in der Südost-Sahara: Erdkunde, v. 22, p. 257-269.

_____, and Pachur, H.J., 1971, Observations on climatic geomorphology and Quaternary evolution of landforms in south central Libya: in, Gray, Carlyle, (ed.), Sym-posium on the Geology of Libya: Tripoli, 1969, Faculty of Science, Univ. of Libya, p. 387-400.

Harger, H.S., 1914, Some features associated with the de-nudation of the South African continent: Geol. Soc. South Africa Proc., Trans. v. 16, p. 22-41.

267

Hedin, Sven, 1903, Central Asia and Tibet: v. 1 and 2, Charles Scribners & Sons, New York, 608p.

_____, 1905, Journey in Central Asia 1899-1902: Lithographic Inst., Gen. Staff Swedish Army, Stockholm, Sweden, 1,241p.

Kaiser, Erich, 1926, Die diamantewüste Sudwest-Afrikas: Dietrich Reimer, Berlin, v. 2, p. 214-339.

Keyes, C.R., 1909, Erosional origin of the Great Basin Range: Jour. Geol., v. 17, p. 31-37.

Klitzsch, Eberhard, 1966, Comments on the geology of the central parts of southern Libya and northern Chad: in, Williams, J.J., (ed.), South-Central Libya and Northern Chad: 8th Ann. Field Conf., Guidebook, Petroleum Expolor. Soc. of Libya, p. 1-17.

Krinsley, D.B., 1970, A geomorphological and paleoclimatological study of the playas of Iran: U.S. Geol. Survey Final Sci. Rept., Contract No. PRO CP 70-800, 2v. 486p.

McCauley, J.F., Grolier, M.J., and Breed, C.S., 1977a, Yardangs of Peru and other desert regions: U.S. Geol. Survey Interagency Report: Astrogeology, v. 81, 176p.

_____, Ward, A.W., Breed, C.S., Grolier, M.J., and Greeley, Ronald, 1977b, Experimental modeling of wind erosion forms: Reports of Accomplishments of Planetology Programs, 1976-1977, NASA TM X-3511, p. 150-152.

Mainguet, Monique, 1968, Le Borkou, aspects d'un modelé eolien: Ann. de Géographie, v. 77, p. 296-322.

_____, 1970, Un etonnant paysage: Les cannelures gréseuses du Bémbeche (N. du Tchad). Essai d'explication géomorphologique: Ann. de Géographie, v. 79, p. 58-66.

_____, 1972, Le modelé des grés: Problèmes généraux Inst. Geog. National, Paris, 2 v., 657p.

_____, Callot, Y., and Guy, Max, 1974, Systemé Cretês-Couloirs: Photo-interpretation, v. 13, no. 74-1, Ed. Technip, Paris, p. 24-30.

ONERN, 1971a, Cuenca del Rio Pisco: in, Inventario, evaluacion y uso racional de los recursos naturales de La Costa: Oficina Nacional de Evaluacion de Recursos Naurales (ONERN), Republica del Perú, Jan., v. 1, p. 36-76, v. 2, p. 2-12, 90.

ONERN, 1971b, Cuenca del Rio Ica: in, Inventario, evaluacion
y uso racional de los recursos naturales de La Costa:
Oficina Nacional de Evaluacion de Recursos Naturales
(ONERN), Republica del Perú, May, v. 1, p. 41-81, v. 2,
p. 1.21.

Peel, R.F., 1970, Landscape sculpture by wind: Selected
papers, v. 1, Physical geography, 21st Internat. Geog.
Cong., Calcutta, India, p. 99-104.

Smith, H.T.U., 1963, Deflation basin in the coastal area of
the Sechura Desert, northern Peru: Tech. Rept. 3-A,
Office of Naval Research, Geog. Branch, Univ. Mass.,
Amherst, 16p.

Thompson, D.G., 1929, The Mohave Desert region, California:
U.S. Geol. Survey Water Supply Paper 578, 759p.

Tricart, J., and Cailleux, A., 1969, Le modelé des régions
sèches: Traité de Géomorphologie, Sociéte d'Edition
d' Enseignement Supérieur, Paris, p. 307-309.

Walther, Johannes, 1924, 1924, Das Gesetz der Wüstenbildung
in Gegenwart und Vorzeit: Quelle und Meyer Verlag,
Leipzig, 4th ed., 421p.

Worrall, G.A., 1974, Observations on some wind-formed
features in the southern Sahara: Zeitschr. fur
Geomorph., N.F., v. 18, p. 291-302.

THE MARIA EFFECT: EQUILIBRIUM AND ACTIVATION OF AEOLIAN

PROCESSES IN THE GREAT BASIN OF NEVADA

Wilton N. Melhorn
Department of Geosciences
Purdue University

Dennis T. Trexler
Nevada Bureau of Mines & Geology

ABSTRACT

The Great Basin of Nevada has abundant small sand dunes and other aeolian features leeward of most intermittent lakes or major drainage courses. The greatest volume, however, presently is lodged as thin sand sheets stabilized by desert shrubs. Other bedforms, such as ubiquitous shrub coppices and minor clay dunes are small and also quasistable under the present climatic and vegetational regime.

Large individual dunes or dune complexes are uncommon but respond to: (1) geographic localization that results from a single "point source" of sand, creating an envelope of downwind transport and deposition; (2) reactivation of stabilized sand sheets or dune surfaces by removal of anchoring vegetation preparatory to irrigated agriculture in desert basins and (3) diminution of sand supply as engineering works progressively entrap or decrease sediment formerly available for transport. Examples of the respective types are: (1) Crescent Dunes near Tonopah; (2) Big Dune near Beatty, and Desert Valley northwest of Winnemucca and (3) the Sand Mountain-Walker River area east and south of Fallon, Nevada. A fourth example, in Clayton Valley near Silver Peak, may represent ongoing changes in dune position and form resulting from an unusual combination of causal factors, a mining induced lowering of the water table and an incomplete adjustment to changes in postglacial wind circulation.

271

Repetitive movement of salt crystals, silt, and clay by "dust-devils" is relatively insignificant and ceases when deflation reaches a resistant substrate. Presently active dune sand is derived from remobilization of preexisting dune forms, and where disturbed by man's activities, movement is sufficient to endanger highways and range land or greatly increase soil erosion. Increased agribusiness and installation of deep-well irrigation systems in desert valleys where electric power is just now becoming available results in a threshold crossing that is reactivating stabilized dunes or sand sheets.

. . . And they call the wind Maria . . .

Milton Keynes UK
Ingram Content Group UK Ltd.
UKHW040443071024
449327UK00020B/954